I0277136

UNITED STATES CRYPTOLOGIC HISTORY

Series 1
Pre-World War I
Volume 1

Masked Dispatches: Cryptograms and Cryptology in American History, 1775-1900

Ralph E. Weber

CENTER FOR CRYPTOLOGIC HISTORY

NATIONAL SECURITY AGENCY

First Edition: 1993
Second Edition: 2002

Table of Contents

Foreword

This is an examination of codes and ciphers as they figured in American history prior to the twentieth century, prior to the era of wireless or radio communication and the advent of the electronic age. It forms a backdrop for understanding modern cryptology and the role of cryptology (notwithstanding its traditional secrecy) in the growth of this nation. Our guide is Dr. Ralph E. Weber of Marquette University, whose 1979 *United States Diplomatic Codes and Ciphers, 1775–1938* (Chicago: Precedent Publishing Inc.) established him in the forefront of students of this arcane subject.

Cryptology, the art and science of code-making (cryptography) and code-breaking (cryptanalysis), depends on the prevailing state of technology and the perception of threat:

• Technology determines the means of communications. Technology also provides the means for protecting and the means of exploiting intercepted communications.

• Perception of threat depends upon a number of considerations, such as the estimated degree of risk, or the damage that might occur, should an unintended recipient become privy to the contents of the communication.

The perception of threat rises naturally in war, but it also pertains in international relations, in business competition, in politics, and even in personal matters, including financial transactions. Applying technology to protect communications (to "mask" them, to use Thomas Jefferson's term) or to exploit those of another party introduces other variables, not the least of which is cost. Cost can involve dollars of time, including personal inconvenience. Sometimes the risk is discounted, if the cost seems too great; conversely, faced with the consequences of compromised information, what seemed earlier too great a cost may well be dismissed in favor of security.

America was born out of revolutionary conspiracy. One of the principal concerns for conspirators is communication, keeping in touch, and doing so in confidence. As rebels and conspirators, the young nation's leaders had turned to codes and ciphers in an effort to preserve the confidentiality of their communications. The *technology* of the time was that of messenger or hand-written correspondence, hand delivered, or by prearranged signals, such as Revere's fabled lanterns, "one if by land, two if by sea." The *risk* was that a dispatch might fall into enemy hands through capture of a courier, and this did happen. Intercepted cryptograms yielded to cryptanalysis of an elementary sort, producing communication intelligence, COMINT. But there was no COMINT effort as we would understand the term today. The technology of the time would not have supported such a concept. (How many enemy couriers could be scheduled for systematic and regular capture, to justify thought of a sustained effort?) Nor was cryptography an organized bureaucracy; rather, it depended upon the interest, knowledge, and imagination of a few men. America was lucky to have such men when it needed them. Some were "civilians in uniform," volunteer soldiers; most were learned men, clergy (familiar with Greek, Latin, Hebrew), mathematicians, scholars — some were statesmen. Their involvement in cryptology was generally brief, but it constituted the seeds of American cryptology.

With the successful end of the Revolutionary War, the occasion for COMINT disappeared, as did the perceived need for secrecy, in the absence of an adversary. When the need subsequently arose, particularly in the case of foreign affairs – when knowledge of the plans and actions, strengths and weaknesses, of one party by another could well thwart the young republic – it was natural to return to forms of cryptography recalled from the Revolutionary experience: the dictionary code, nomenclators, simple cipher.

No formal documentation of official American government cryptography is known to have been recorded – indeed, governments and regimes have traditionally been reluctant to publicize such activities. There is little evidence that American leaders were acquainted with what we now view as the classics on the subject of cryptography, although exceptions cannot be dismissed. Historically, what was learned through cryptanalysis of another's communications has had an effect on one's own cryptography. One might speculate that what was learned through experience by government clerks or officials was passed on through "on-the-job training" and occasional notes of instruction, and some of the latter do exist. Around the turn of the eighteenth/nineteenth century, an impressive treatise of some 35,000 words was generated by an English surgeon for an encyclopedia. Written around 1807, it was another decade before Dr. William Blair's article "Cipher" appeared in a volume of the serially-issued *Cyclopedia,* edited by Abraham Rees in London, and an American edition subsequently appeared. In his *The Codebreakers* (1967), David Kahn rightly characterized the Blair piece as "the finest treatise in English on cryptology," until army lieutenant Parker Hitt's little military manual appeared in 1916. Blair distilled the essence of the art and science of cryptology from the ancients (along with an impressive bibliography for the time) and offered the results of his own study and deductions. Surely the encyclopedia piece must have been read by some in government service, yet no firm evidence of the fact, or of its influence, has been found, apart from its having been used as a training manual by the army's signal corps (or Signal Service) in the post-Civil War years. (See chapter 14.) Perhaps it would be expecting too much to find it cited (although it is known to have been a source for Edgar Allan Poe, who, in turn, is cited by others).

Dr. Weber's examples also show the influence of communication technology – the postal card, the telegraph, and the transatlantic cable – on American cryptography. Cost becomes a more dominant consideration for the government than security when electrical communications are introduced (perhaps with the exception of the Civil War years, when the War Department set up its own U.S. Military Telegraph). Manual or mechanical devices began to appear in connection with cryptography – Jefferson's well-known "wheel" or cylinder cipher is an example. By the end of the period, mechanical devices have become more sophisticated, and the age of electromechanical devices and machines is just ahead.

David W. Gaddy

Introduction

But why a cypher between us . . . there may be matters merely personal to ourselves, and which require the cover of a cypher more than those of any other character. This last purpose, and others which we cannot for[e]see may render it convenient & advantageous to have at hand a mask for whatever may need it.

President Thomas Jefferson to Robert Livingston, 1802

Almost two months after the declaration of war against Spain in 1898, the following coded dispatch, cabled to the secretary of state from the American minister in Madrid, arrived at the White House in Washington, D.C.:

CABLE. .2 W Z SX 115 Govt...PARIS Via French. June 13th '98-550p
DAY, Secretary Washington.

Is [*sic*] my confidential letter seventh
41502 22417 37577 of the 48994 instant hy [*sic*]

I fully explain minister for foreign
english mail 23790 the wish of 35793

affairs [Mr. Hanotaux] bring about treaty of peace
to 19084 [15084] 42628

his willingness to lend unofficially
and 37334 55468 30177 himself 58145 [57145] and in

a friendly action which may lead
37709 manner to any desired 18807 37988 32075 to that

result says he can the [*sic*] ambassador
46077 he now 40986 [48986] 13990 place the 17625

in contact with madrid if I am authorized to talk
here 22224 38084 [33184] 37040 12858 51522

with him he armistice will be
57169 on the subject 32631 supposed 1911242[*sic*] 50867

suggested asked means of opening negotiations
59715 or 18948 as a 31480 [31489] 46418 43406 but

cannot be conceded spain believes
that of course 16491 26114 now 53804 he 12675

is ready for peace she may save
48069 [48060] making 41628 now when 42396 32688 still 43786

something my letter telegraph if I can
55503 after reading 37677 please 52923 me 16990

consent to see spanish ambassador how far
24220 40990 the 16625 and 26296 [26995]

I go in discussing the subject with him
30039 can 30316 25262 52912 57169 without

making commital [*sic*]
37485 any 25509

PORTER [1]

This coded telegram to the newly installed secretary of state, William R. Day, from General Horace Porter, the American ambassador to France, received in the Executive Mansion at 5:50 P.M. June 13, was decoded, and a plaintext copy promptly sent to an anxious President William McKinley at 7 P.M.[2] Transmitted from a European nation long famous for intercepting foreign dispatches and breaking secret codes, the cablegram masked the discussions that a tactful Porter, West Point graduate, former military secretary to President Ulysses Grant, and later energetic railroad executive, had held with the French foreign minister, Gabriel Hanotaux. Written less than two months after the beginning of the Spanish-American War, the secret dispatch revealed French willingness and support for arranging an armistice between weakened Spain and the United States as a prelude to a treaty of peace. And between its lines, it also told a worried McKinley and Day that French hostility toward the United States was lessening as an anxious France sought to maintain neutrality in the armed conflict.

The confidential cable also portrays the State Department's cryptological sophistication by the late nineteenth century. The five-digit codenumbers, based on The Cipher of the Department of State, published in 1876, graphically mirror the code clerk's process of masking clandestine messages between the department and American ambassadors overseas. In a special effort to improve communications security, the department modified the codenumbers in the book according to the following pattern: in the five-digit number, the first digit stayed in place; the last two digits were reversed and moved to second and third place; the two digits that had been in second and third place were moved, respectively, to fourth and fifth place.[3] This dispatch also illustrates multiple mistakes in either encoding and/or transmitting the cable, and the resulting uncertainties in the important message because of these errors.

Because of the Atlantic cable, opened in 1866, dispatches from Europe reached the White House in minutes rather than the three to four weeks and more that occurred earlier in the century. This stunning revolution in electromagnetic communications reduced transmission time and empowered modern American presidents and secretaries of state to secure timely information, and thus function as better informed and often more effective commanders and diplomats.

Decisive for the security process were the secret codes that maintained the communications security absolutely crucial for effective negotiations.

By the end of the nineteenth century, American confidential communications systems for foreign correspondence had advanced well beyond the simple cipher and code systems first employed in 1775. If the telegram above had been veiled in America's very first code, the United Colonies code, the first sentence of that dispatch would have read as follows:

> 16, 12, 76, 40, 49, 105, 54, 23, 26, 86, 97, 174, 19, 126, of the 117, 13, 121, 116, 51, 97, 613, [by] English mail I 655, 600, 631, 623, 643 40,the wish of 7621, 63, 661, 87, 142, 59, 157, 199, 196, 231, 199, 253, 216, 240, 203, 472, 406, 378, 395, 424, 387, 551, 411, 413, to 259, 224, 191, 311, 561, 440, 500, 556, 557, 555, 389, 398, 397, 430, 490, 40, 67, 23, 60, 59, 19, 28, 26, and 613, 601, 689, 1, 8, 66, 18, 73, 21, 27, 238, 229, 217, 215, 357, 315, 253, 220, 217, 229, 289, himself 350, 342, 353, 429, 600, 573, 591, 655, 593, 657, 663, 40, and in 593, 600, 592, 601, 604, 602, 603, 643, 40, manner to any desired 378, 423, 521, 546, 545, 406, 472, 472, 613, 601, 439, 613, 420, 378, 40, 508, 491, 471, 451, to that 411, 399, 400, 404, 466, 436, 83

By 1900, American secret systems for veiling messages had become more effective and efficient; moreover, certain attitudes towards secret writing were gradually undergoing a remarkable conversion. John Haswell, the post-Civil War State Department codemaker, described this significant transformation in the late nineteenth century when he wrote with a mixture of naivete and exaggeration: "In former times they [ciphers] were employed for purposes of evil and cruelty, and were consequently looked upon with horror and aversion."[4] However, Haswell thought optimistically, "Their functions now, however, are chiefly to benefit humanity by facilitating commerce and industry, and hence they merit public interest and favor." Not altogether accurately, Haswell believed that previously ciphers had been mainly a "war factor" and had been incorporated in the military systems throughout the world. Since the invention of the telegraph, however, he thought that cipher operations had followed peaceful pursuits, and they had become essential elements in all financial, commercial, and industrial establishments. And for some anxious observers, the advent of telegraphic communications "baptized" the study and practice of secret writing in peacetime, and lessened the suspicions about those persons engaged in this questionable discipline.

At the time of the American Revolution, the American Founding Fathers did not believe codes and ciphers "were employed for purposes of evil and cruelty." Rather, they viewed secret writing as an essential instrument for protecting critical information in wartime, as well as in peacetime. The newly established American nation, formed from financially distressed and war-ravaged states, was a weak union, struggling in its early decades with daily internal stresses. Revenue taxes, domestic trade agreements, and political dissension threatened the new government as did foreign enemies and intrigues. Because the fledgling nation frequently distrusted the motives of European diplomats, and properly feared the machinations, international political alliances, and traditional European practices such as espionage, the Founding Fathers quickly, though sometimes inexpertly, recognized the dire necessity for more secure communications. Early in the American Revolution, and especially evident during the military struggles in the fog of war, communications security became a decisive objective. The frequent interception of American diplomatic correspondence by "black chamber" agents in European capitals taught prominent American statesmen to create cipher and code systems for confidential written correspondence. Only with this arrangement could the fledgling and beleaguered United States hope to conduct diplomatic negotiations with confidence and success.

Charles W. F. Dumas, James Lovell, William Lee, John Jay, Francis Dana, Thomas Jefferson, James Madison, James Monroe, John Quincy Adams, the Marquis de Lafayette, Robert R. Livingston, Major Benjamin Tallmadge, Robert Morris, Charles Thomson, General George Washington, Edmund Randolph, Alexander Hamilton, Oliver Wolcott, Aaron Burr, William Vans Murray, Robert Patterson, Nicholas Trist, General Albert Myer, Anson Stager, and John H. Haswell created and/or encouraged the American use of codes and ciphers for communications security between 1775 and 1900. These distinguished statesmen and talented public figures realized the dire necessity for secret correspondence and urged the Congress and their associates in various federal government offices to develop and employ confidential communiques. And with the development of the telegraph and cable, it became even more evident that the transmission of classified information over these public communication lines demanded encrypted systems.

At least five American names figure prominently in the development of communications intelligence during the century after the Battle of Breed's Hill: Colonel Elbridge Gerry, Elisha Porter, Reverend Samuel West, James Lovell, and Charles A. Keefer. The first three individuals managed to unlock the secret monoalphabetic substitution cipher found in a carefully disguised and detailed report that was intercepted on its way from the director and physician of the first Continental Army hospital, poet and traitor Dr. Benjamin Church, to a British major, Maurice Cane, in Boston. The codebreakers provided General George Washington with the vital evidence that led to Church's imprisonment for spying in the fall of 1775 and a sentence of exile in the West Indies.

James Lovell also had great success in breaking enciphered dispatches from General Henry Clinton to Lord Cornwallis, which were intercepted in 1780 and 1781. And decades later, Charles A. Keefer, a cipher clerk and civilian telegrapher for General Philip Sheridan in New Orleans in 1866, intercepted highly significant French dispatches being transmitted from Mexico City via New Orleans to Paris, and from Paris to Mexico City. Keefer is probably the first person in the service of the United States to use communications intelligence in peacetime. It is also important to note that biographies of Gerry, West, and Lovell are published in the *Dictionary of American Biography*; however, no mention is made of their codebreaking activities.

By the mid-eighteenth century, European intelligence officers had skillfully developed cryptographic designs that included complicated and efficient cipher and code systems. Moreover, the skills exhibited in Vienna, London, Paris, and Madrid "Black Chambers" for intercepting foreign and domestic dispatches and breaking cryptographic systems had matured after generations of study and practice. The traditions of cryptography established in Western Europe moved slowly to the United States in the period after 1775. In the crucible of war, desperate American leaders struggled to learn the ways and means of secret correspondence. Surprisingly, in contrast to European practices, there were apparently no peacetime professional codebreakers in the United States until after the World War I period.

An embryonic and besieged United States in 1775 lacked the sophistication, skills, and European diplomatic traditions so integral for successful secret communications systems. Also a certain naivete colored American views. Particular American leaders, buoyed by the theories of the French philosophes, called for a rule of reason and openness in American diplomatic practices. Even John Adams echoed this attitude when he lectured the Compte de Vergennes, the astute and probably surprised French foreign minister:

"The dignity of North America does not consist in diplomatic ceremonials or any of the subtleties of etiquette; it consists solely in reason, justice, truth, the rights of mankind and the interests of the nations of Europe."[5]

Revolutions, however, provide fertile soil for intrigue, espionage, and, of course, secret communications. The Continental Congress recognized the need for secrecy when it passed the following resolution: "If an original page is of such a nature as cannot be safely transmitted without cyphers, a copy in cyphers, signed by the Secretary for the department of foreign affairs, shall be considered as authentic, and the ministers of the United States at foreign courts may govern themselves thereby, in the like manner as if originals had been transmitted."[6] The young American government appreciated the critical need for secrecy in an imperfect world of confidential diplomacy and spying. Gradually, during the Revolution, and sometimes reluctantly, United States leaders acquired some of the cryptographic talents and skills of their European ally France and their powerful enemy Great Britain.

In the early decades, American codemakers, with one major exception, offered nothing innovative for the world of secret communication with their ciphers, book codes, and codesheets. They did eventually, however, establish the system used for State Department secret communications until 1867 through the development of the 1,700-item codesheets, such as the version devised for James Monroe for his negotiations regarding the Louisiana Purchase. The exceptional codemaker was Thomas Jefferson, whose cipher cylinder, which he called his "wheel cypher," offered a brilliant mask, indeed twentieth century security, for secret messages.

The French Revolution, beginning in 1789, increased foreign policy tensions for the United States. Secret messages from American ministers, including those involved in the XYZ Affair, poured into the State Department as American presidents and diplomats sought neutrality and careful isolation from the European conflicts. The first decade of the nineteenth century found the United States as a nervous spectator of the Napoleonic Wars. However, neutral rights at sea and craving for land expansion caused the nation to stumble into what came to be known as the War of 1812. Once again, the United States faced an awesome armed conflict with Great Britain. Diplomatic dispatches before and during the war were veiled with the 1,700-element codesheets devised earlier.

But not all secret messages in America during the decades of the late eighteenth and all of the nineteenth centuries concerned foreign threats of war or ongoing negotiations for opening neutral commerce. In 1764, Thomas Jefferson used a book code to hide the name of a young lady whom he planned to court. And in the years after 1780, Jefferson, James Madison, James Monroe, and a covey of other political leaders in the United States often wrote in code in order to protect their personal views on tense domestic issues confronting the American nation. Employing many codes and a few ciphers, they sought safety for their dispatches: they built security fences to protect their correspondence from political rivals and American postal officials.

Jefferson and Monroe during diplomatic service in France, and John Adams in England, became even more sensitive to the dangers of intercepted dispatches. They carried these experiences and anxieties back to America. Madison, as secretary of state in Jefferson's administration, also acquired valuable insights and further understanding about the necessity to mask messages, domestic and foreign.

Early in the nineteenth century, Aaron Burr, with his plans for expansion and empire, carefully shielded his many designs and instructions in a combination of codes and ciphers for his abortive expedition to the Southwest. His strong desire for acquiring territory from Spanish lands, and perhaps from existing United States territory, brought him to the edge of treason.

In the decades after 1815, probably fewer encrypted domestic messages were carried along the more secure American postal routes. These same decades witnessed a marked decline in encoded American dispatches for the American legations in Europe. Beginning in the mid-1820s, encoded dispatches to and from American diplomats in unstable Mexico City became the most numerous. Most of them were encoded in the same design first used by James Monroe in France. And while many other American diplomats served more as reporters of, rather than actors in, diplomacy, Joel Poinsett and Anthony Butler, American ministers in Mexico, sought to manipulate their host nation's politics and acquire additional Mexican territory for the United States. Their numerous secret encrypted dispatches to Henry Clay, Martin Van Buren, Edward Livingston, and Louis Mcbane reflected the turmoil and numerous clandestine activities of American foreign relations with its southern neighbor.

In the 1840s an American public became fascinated with cryptology because of Edgar Allan Poe's fourth book, *The Narrative of Arthur Gordon Pym*, an imaginative story about South Seas adventures, shipwreck, and the use of cryptograms. His prize-winning story "The Gold Bug," published in 1843, heightened the mystery and magic of secret writing for many readers, young and old. Also, the development of postcards in the 1840s renewed the public's special fascination with confidential writing as correspondents sought to veil personal information from postal carriers.

An American revolution in communication technology began in the 1840s with the introduction of the electromagnetic telegraph, soon to be followed by the development of a transatlantic cable. These exciting and innovative years also witnessed a heavy torrent of secret writing techniques to be used with the magnificent telegraph machine, which reduced communi-cations times from months and weeks to hours and minutes. Thousands of business and diplomatic correspondents soon developed special code and cipher systems, designed for economy and secrecy. And the federal government became increasingly involved in communications security, especially during wartime.

In 1871 William Whiting, the assistant to the attorney general, recognized the crucial importance of the telegraph: "In time of war, the lines of telegraph have now become as indispensable as arms and ammunition. By their agency, the Government becomes omnipresent, and its powers are immeasurably enhanced. The movement of armies and navies are controlled, life and property are protected and the voice of authority uttered at the Capitol is heard almost instantaneously throughout the country."[7]

In the mid-1850s, British military forces in the Crimea used the telegraph for strategic lines; however, operational and mobile use of that instrument began during America's Civil War. Shortly before that war began, a brilliant new communications system, a visual system, designed by assistant army surgeon Albert James Myer, became a crucial military companion to Morse telegraph. As David W. Gaddy evaluates so accurately, "Wiretapping, signal interception and exploitation, authentication systems, the 'war of wits' between 'code-making and code-breaking' for Americans truly stemmed from the American Civil War, and Myer's system, as well as the organizational concept of a corps of trained communicators, that made an impact on other armies of the world."

In the Confederacy, simple ciphers, codenames, and book codes (often called at the time "dictionary ciphers" since dictionaries conveniently provided sufficient vocabulary) prevailed, until the poly-alphabetic Vigenère cipher became the standard for government and military. In the North, a route or word transposition system was used to protect the military telegrams transmitted by the U.S. Military Telegraph under the personal control of the secretary of war.

As America turned outward more energetically in its foreign relations in the generation after the Civil War, its leadership still remained reluctant to recognize foreign espionage activities, especially in communications intelligence. This activity, sometimes initiated, at other times renewed, by European nations expanded greatly as these governments supervised telegraph and cable companies. The State Department finally abandoned the Monroe code in 1867 and replaced it with a disastrous codebook, designed for economy and resulting in confusion. Within a decade, a new codebook, *The Cipher of the Department of State*, prepared by John Haswell, replaced the poorly designed book. The new volume established a pattern that remained the basic standard for the department's communications for the next two generations.

In the United States, the 1880s witnessed the formation of an energetic new navy, with cruisers of steel and with more powerful guns, developed in the Washington Navy Yard: construction of a fleet of battleships commenced in 1890. Also, the navy and army gradually established intelligence offices: the Office of Naval Intelligence in 1882 and three years later the Army's Military Information Division. In 1885, the navy published *The Secret Code*, a massive book, with two supplemental volumes, which incorporated the superencipherment of code. At this same time, the War Department published its first telegraph codebook, replacing postwar versions of the route transposition system used in the 1870s. Thus, more secure American military communications systems were in place by the beginning of the Spanish-American War.

Looking back on the first four generations of masking American dispatches, 1775 to 1900, it is very evident that the Founding Fathers were much more anxious than their successors to encrypt their confidential correspondence. With the exception of George Washington, all the presidents before Andrew Jackson had served overseas or had functioned as secretaries of state and thus were exceptionally sensitive to European, and sometimes domestic, interference in written communications. However, for the secretaries of state and presidents after Jackson, key concepts in masking dispatches were thriftiness and tradition, not security. Thus, the State Department practice of using the Monroe Code for over sixty-three years and also the ill-designed 1867 Code are the best examples of this misguided rationale.

The relatively lax practices concerning the 1876 Code provide additional evidence of the failure to understand foreign interception and intelligence customs.

The exciting technology of the electromagnetic telegraph and the telegraph cable spurred the surge of secret writing and codebreaking in the Western world. After a short time, the techniques also hastened the creation of more sophisticated codes and ciphers. It is abundantly evident that secret communications played a fundamental role in American foreign relations, in peace and war, from the XYZ Affair to the intercepted Spanish cables in 1898.

Much less is known about U. S. codebreakers during these four generations. Apart from the Civil War efforts of David Homer Bates and his colleagues in the War Department telegraph office, the splendid contributions of Elbridge Gerry, James Lovell, and Charles Keefer in the service of their country, and private efforts by John Hassard, William Grosvenor, and Edward Holden, highlight the history of this challenging and fascinating science. It remained for twentieth-century America, confronting powerful foreign enemies in World War I, and shocked by a devastating enemy attack on Pearl Harbor, to recognize finally the crucial necessity for maintaining secure communications and obtaining foreign intelligence data during war and peace.

While our third president, Thomas Jefferson, wrote about the necessity of masking dispatches, a recent president, Ronald Reagan, reminded us about the crucial role of American history, a role that is especially applicable to cryptologic studies: "If we forget what we were, we won't know who we are." And then he added, "I am warning of an eradication of the American memory that could result, ultimately, in an erosion of the American spirit."

In preparing this cryptologic study of American history, I was fortunate to receive generous support from many persons who provided me with special assistance, insights, and knowledge about the American cryptologic past. I am especially grateful to David W. Gaddy, whose essays, encouragement, and enthusiasm have enhanced this manuscript. In similar fashion, Henry F. Schorreck offered extremely knowledgeable comments about this pioneering era in American history. I welcomed Barry D. Carleen's editorial advice and David A. Hatch's answers to my numerous queries concerning communications security. Donald M. Gish supplied helpful insights on cryptanalysis; and Robert N. Spore offered valuable computer support and assistance. These splendid persons and the following talented associates and staff members of the Center for Cryptologic History made the complex research and writing of this cryptologic study a special pleasure: Charles W. Baker, Thomas L. Burns, Earl J. Coates, Vera R. Filby, Jules Gallo, Joyce Hamill, Jack E. Ingram, Christina Kikkert, Robert E. Newton, Frederick D. Parker, Jean M. Persinger, Donald K. Snyder, and Guy Vanderpool. In addition, Michael L. Peterson's essay on the Church Cryptogram brought distinctive insights to early American communication security and foreign intelligence.

For their gracious assistance in locating various American manuscript sources, I wish to express my gratitude to John McDonough, Michael Klein, and Nancy Wynn at the Library of Congress; Frank Burch, Jennifer Songster-Burnett, John Butler, Robert Coren, Milton Gustafson, J. Dane Hartgrove, Wilbur

Mahoney, Larry McDonald, Michael Musick, Katherine Nicastro, David Pfeiffer, William Sherman, and John Taylor at the National Archives; Louis Fine and Bruce Kennedy at Georgetown University; and Karl Kabelac at the University of Rochester.

Chapter 1

United Colonies' Cipher

Early in the violence of the American Revolution in 1775, the United Colonies, as they styled themselves before the Declaration of Independence, appealed to King George III for substantial British colonial reforms. As the April violence at Concord and Lexington spread to Breed's Hill in Boston and to New York, patriot leaders, frustrated by the king's inertia, Parliament's continued repression, and the early successes of British troops, sought secret aid from France. Moreover, applauding the appeals of the Philadelphia political pamphleteer Thomas Paine in his essay "Common Sense," published in January 1776, the colonies increasingly united for independence.

In 1775 Charles William Frederic Dumas, one of America's first secret agents, designed and dispatched the first revolutionary secret diplomatic cipher to Benjamin Franklin to mask correspondence between the Continental Congress and its foreign agents in Europe. Dumas, a talented translator and classical scholar, then in his fifties, had been born in Germany, lived in Switzerland, and in 1750 moved to the Netherlands. Earlier, when Franklin was stationed in London, Dumas requested information about settling in America: Franklin advised migrating to New York, New Jersey, or Pennsylvania rather than East Florida, which Dumas had mentioned, because the three established colonies offered more intellectual stimulation, especially for a man of letters. Dumas had translated the celebrated Law of Nations, written by the Swiss scholar and jurist Emerich de Vattel in 1760; moreover, he sent Franklin three copies at the time of the revolution in hopes the legal principles would guide the fledgling nation.

Benjamin Franklin

Dumas, paid by the United States £100, later 200 louis d'ors a year, was given secret assignments requiring him to report on foreign diplomats stationed in Holland, to disseminate propaganda favorable to the United States, and win Dutch support for the American war maneuvers. Well aware of European espionage programs, especially those involving postal inspection and intercepts, Dumas carefully instructed American diplomats about the difficulties they faced. And he also confronted danger, for the British ambassador at The Hague intercepted an incriminating letter to Dumas in 1776 and learned of this American spy.

From the full passage, two lists were prepared: an alphabetical list of letters for enciphering with the different numbers for each letter of the alphabet noted alongside the letter; for deciphering, a list in numerical order was used. The full passage, with 682

The Dumas Cipher (Partial)

v	o	u	l	e	z	-	v	o	u	s	s	e	n	t	i	r
1	2	3	4	5	6	7	8	9	10	11	12	13	14	15	16	17

l	a	d	i	f	f	e	r	e	n	c	e	?	j	e	t	t
18	19	20	21	22	23	24	25	26	27	28	29	30	31	32	33	34

e	z	l	e	s	y	e	u	x	s	u	r	l	e
35	36	37	38	39	40	41	42	43	44	45	46	47	48

symbols, did not include the letter w, and therefore enciphering instructions specified that two u's should be used (although Franklin used two u's). Also k was not included, and c was substituted in the dispatches in the early months. (A subsequent edition was to add four different numbers for k.) When properly employed, the cipher offered modest security because there were 128 different numbers for e, 63 numbers for r, 60 for s, 50 for a, and 44 for o. Unfortunately, persons, including Dumas, who enciphered dispatches tended to use only the first listed numbers for each letter rather than a random selection: this practice, together with the inherent weakness involved in basing a cipher on a paragraph of prose, simplified the codebreaker's assignment. Despite these defects, Franklin, then in his seventies, and stationed as American commissioner in Passy, near Paris, used the cipher as late as 1781 to tell Dumas about new negotiations for a peaceful resolution to the war:

Dear Sir,

We have News here that your Fleet has behaved bravely; I congratulate you upon it most cordially.

I have just received a 14. 5 .3. 10. 28 .2. 76. 202. 66. 11. 12. 273. 50. 14. joining 76. 5. 42. 45. 16. 15. 424. 235. 19. 20. 69, 580. 11. 150. 27. 56. 35. 104. 652. 28. 675. 85. 79. 50. 63. 44. 22. 219. 17. 60. 29. 147. 136. 41. but this is not likely to afford 202. 55. 580. 10. 227. 613. 176. 373. 309. 4. 108. 40. 19. 97. 309 17. 35. 90. 201. 100. 677.

By our last Advices our Affairs were in a pretty good train. I hope we shall have advice of the Expulsion of the English from Virginia.

I am ever,

Dear Sir, Your most obedient &
most humble Servant
B. Franklin[1]

The plain text for the enciphered message paragraph was as follows:

> I have just received a neuu commissjon [*sic*] joining me uuith m adams in negociaions [*sic*] for peace but this is not likely to afford me much employ at present.

Dumas would also employ another one-part cipher sheet in correspondence with Franklin when the latter was stationed in France. This one contained 928 elements and listed words, mainly in alphabetical order, beginning with number 1 and continuing through 923. The weakness inherent in an alphabetical-numerical sequential code was increased by the fact that this particular one contained cipher equivalents for only sixteen out of the twenty-seven most frequently written English words. Despite these deficiencies, this instrument proved more efficient and attractive, especially to Franklin, than the cipher originally introduced by Dumas in 1775.

Notes

1. Franklin to Dumas, 16 August 1781, Benjamin Franklin Collection, Yale University, New Haven, Connecticut.

Chapter 2

"Friend Jimmy's Cyphers": James Lovell and Secret Ciphers during the American Revolution[1]

When the American Revolution began, American statesmen found themselves caught up in a violent conflict that demanded more than guns and powder. For the first time, the new nation faced the international world of intrigue and spies as it negotiated with sovereign nations. Though familiar with the traditional weapons of war, Americans lacked the experience and sophistication required for secret diplomatic correspondence. European countries, particularly Austria, France, and Great Britain, spent large sums of money to develop cryptographic systems for protecting their foreign correspondence. Moreover, they also organized confidential offices to intercept and cryptanalyze the diplomatic correspondence of foreign ministers stationed in their respective countries. These postal intercept and solving agencies, termed "Black Chambers," enabled government officials to read foreign confidential dispatches, frequently before the dispatches were delivered to the proper addressee. Although the United States, unlike other powers, did not introduce an official codebreaking office until the twentieth century, the American revolutionary generation was aware of the dangers of interception and sought to protect the official correspondence of its foreign ministers.

As Congress appointed ministers to European posts, the necessity for secret systems in addition to the Dumas cryptographic design became more evident. Moreover, American leaders also experienced the British seizure of American ships and dispatch pouches, and became familiar with European postal intercept systems. Such actions angered Americans and also led them to the use of invisible ink and a courier system.

A few Americans, most notably James Lovell, sought to counter British, French, and Spanish postal espionage by using ciphers to mask correspondence primarily between Congress and her ministers. Simple cipher designs could be committed to memory and thus promised more security than lengthy code sheets or nomenclators. With his cipher designs, Lovell became America's first cryptographic tutor. Unfortunately, his students, the American ministers abroad, though brilliant and talented in political matters, found his systems confusing and frustrating.

James Lovell, born in 1737, studied at Harvard, taught in his father's school in Boston, and became a famous orator. Arrested by the British after the battle of Breed's Hill, he was sent as a prisoner to Halifax in 1776; but soon thereafter he was exchanged and returned to Boston. Chosen as a delegate to the Continental Congress, he attended the sessions of the Congress beginning in February 1777 and served continuously until the end of January 1782 when he took his only leave. In May 1777, he was appointed to the Committee for Foreign Affairs, where, among other responsibilities, he deciphered dispatches. He became the Committee's most indefatigable member, indeed, sometimes its only active member; other members arrived and departed, but Lovell stayed on and for five years never even visited his wife and children. Before he left Congress in 1782, Lovell had left his mark on American foreign relations and particularly on cryptography.

The Lovell cipher system is based upon the first two or more letters in a keyword. Beginning with each key letter, twenty-seven-item alphabets that include the ampersand are listed. For example, using the first three letters of the keyword "BRADLEY," there are three columns and twenty-seven short rows with the first column ranging from B through & to A; the second, R through Q; the third, A through &. Down the left margin run the numbers 1 through 27.

1	BRA	15	PEO
2	CSB	16	QFP
3	DTC	17	RGQ
4	EUD	18	SHR
5	FVE	19	TIS
6	GWF	20	UJT
7	HXG	21	VKU
8	IYH	22	WLV
9	JZI	23	XMW
10	K&J	24	YNX
11	LAK	25	ZOY
12	MBL	26	&PZ
13	NCM	27	AQ&
14	ODN		

To encipher, the writer finds the first plaintext letter in the column under B and replaces it with the number to its left. He finds the second plaintext letter in the column under R and replaces it with the number of that row. For the third letter, the writer repeats the process with the A column. The cycle is repeated for subsequent letters. A crucial rule in this system provides that when a passage or word is to remain unenciphered, the continuity is broken, and the next word to be enciphered starts its encipherment with the column under B. Sometimes the numbers 28, 29, and 30 are employed as nulls. Occasionally, however, 38 followed by 29 at the beginning of a passage indicates the plain text is enciphered in the normal order of B, R, A; however, 29 followed by 38 at the beginning means encipherment is in reverse order.[2]

James Lovell enjoyed the challenge of making and breaking cipher systems. Soon after John Adams left America for France in November 1777, Lovell wrote to him using CR as the cipher key.

Apparently, Lovell first gave written instructions to John Adams regarding the cipher in May 1780 when he explained — not at all clearly — that the cipher system was based upon the alphabet squared with the key letters being the first two letters of the surname of the family where he and John Adams had spent the evening before going to Baltimore. Lovell's instructions described a column of the alphabet as twenty-six letters beginning with A and including as the twenty-seventh element &. In the next column, Lovell began with B and carried it through the fourth letter; the next column began with C and again listed the alphabet only through the fourth letter. This was the only design he gave Adams for the alphabet squared, and probably this increased the

				b	a	n	k	r	u	p	t	
I can only say that we are				27.	11.	12.	21.	16.	4.	14.	3.	
w	i	t	h	a	m	u	t	i	n	o	u	s
21.	19.	18.	18.	26.	23.	19	.3.	7.	24	.13.	19.	2.
a	r	m	y								d	e
26.	1.	11.	8.	the latter owing very much to the							2.	15.
l	a	y	o	f	c	l	o	a	t	h	i	n
10.	11.	23.	25	4.	13.	10.	25.	26.	3.	6.	19.	12.
g												
17.[3]												

confusion about the cipher. In writing to John Adams in Amsterdam on 21 June 1781, Lovell again used the cipher key CR, which he explained again guardedly in his letter of 30 November 1781: "You certainly can recollect the Name of that Family where you and I spent our last Evening with your Lady before we sat [*sic*] out on our Journey hither. Make regular Alphabets in number equal to the first Sixth Part of that Family name."[4] The name Lovell alludes to is CRANCH; however, he erred when he said a sixth part, and instead he should have written the "first third part." As a result of such mistakes, John Adams had many problems with enciphered messages, partially because he did not completely understand the design but also because of enciphering errors.[5]

In a letter to Abigail Adams of 19 December 1780, Lovell explained the necessity for a cipher and sought to convince her of its value. He stated that he did not want any of his letters to Adams thrown overboard, as was the custom when a packet ship was in danger of capture by the enemy, unless he specified on the cover of the letter that it was to be thrown overboard. A confident Lovell stated that his cipher would protect the message: "I chalenge [*sic*] anybody to tell the Contents truly. . . I am told the Enemy have another Mail of ours or yours, this prevents my giving you such Explanations of my private letter to Mr. A as I at first intended."[6] He chided Abigail about being averse both to ciphers and to his enigmatic character. Had she felt otherwise about ciphers, Lovell wrote, "I would have long ago enabled you to tell Mr. A some Things which you have most probably omitted." He promised to send her a key to use on special occasions in letters to or from her husband. Lovell again implied that the use of a cipher would save the letters from being thrown overboard if the vessel were in danger of capture: "I am told Letters from Holland have been thrown from Vessels now arrived at Boston when only chased. Those losses at least might be avoided."

Despite Lovell's frequent urging, there is no evidence that John or Abigail Adams ever enciphered letters in the Lovell design; moreover, both had great difficulty in deciphering messages using Lovell's creation.[7] Abigail thanked Lovell in June 1780 for the alphabetical cipher that was sent to her but thought she would never use it: "I hate a cipher of any kind and have been so much more used to deal in realities with those I love, that I should make a miserable proficiency in modes and figures."[8] She added that her husband held similar views: "Besides my Friend is no [*sic*] adept in investigating ciphers and hates to be puzzled [*sic*] for a meaning." But she did try to decipher the parts of Lovell's letters to her that were written in cipher, as well as copies of Lovell's letters to her husband John. She continued to find the cipher troublesome, though she became somewhat adept in deciphering, mainly due to the help of her friend, Richard Cranch, and only after Lovell in one letter reminded her that the family in the "Evening" referred to Cranch.[9] And, in fact, she instructed her husband, though not very clearly, as to how to use the cipher.

As late as June 1782, almost two and one-half years after the cipher's introduction, she wrote, "With regard to the cypher of which you complain, I have always been fortunate enough to succeed with it. Take the two Letters for which the figure stands and place one under the other through the whole Sentance [*sic*], and then try the upper Line with the under, or the under with the upper, always remembering, if one letter answers, that directly above or below must omitted, and sometimes several must be skiped [*sic*] over."[10] She wrote these words to her husband after reading his complaint in a letter of 12 February 1782, to Robert Livingston: "I know very well the name of the family where I spent the Evening with my worthy friend Mr. [blank space in the original, apparently for security] before we set off, and have made my alphabet accordingly; but I am on this occasion, as on all others hitherto, unable to comprehend the sense of the passages in cypher. The cypher is certainly not taken regularly under the two first letters of that name. I have been able sometimes to decypher words enough to show that I have the letters right; but, upon the whole, I can make nothing of it, which I regret very much upon this occasion, as I suppose the cyphers are a very material part of the letter."[11]

All in all, for John Adams, the Lovell ciphers caused boundless confusion. As Adams confided in a letter to Francis Dana in Paris in March 1781: "I have letters from the President and from Lovell, the last unintelligible, in ciphers, but inexplicable by his own cipher; some dismal ditty about my letters of 26th of July, I know not what."[12] This in spite of Lovell's many explanations, as in his letter of June 1781: "I suspect that you did not before understand it from my not having said supped in Braintree. I guess I said New England."[13] Adams could not read Lovell's enciphered dispatches. Indeed, the instructions to John Adams, Benjamin Franklin, John Jay, Henry Laurens and Thomas Jefferson, ministers plenipotentiary in behalf of the United States to negotiate a treaty of peace, sent after 15 June 1781, by the president of Congress, Samuel Huntington, were enciphered in the CR cipher that Adams found unreadable.

Undoubtedly, Adams must have been pleased with part of Livingston's reply of 30 May 1782, to Adams's letter of 21 February. The reply was a model of courtesy as Livingston apologized for the difficulty that the cipher caused and explained, "It was one found in the office and is very incomplete. I enclose one that you will find easy in the practice and will therefore write with freedom directing that your letter be not sunk in case of danger... want of time reduces me to send you a set of blanks for Mr. Dana which you will oblige me by having filled up from yours with the same Cyphers, and transmitted by a careful hand to him. This will make one cypher common to all three."[14]

After all his uncertainties and difficulties with ciphers, one can only imagine Adams's frustration when he realized Livingston had neglected to enclose the code with the duplicate of his letter. However, Adams simply mentioned this omission in his report to Livingston: "The cipher was not put up in this duplicate, and I suppose the original is gone on to Mr. Dana in a letter I transmitted him from you some time ago, so that I should be obliged to you for another of the same part."[15]

Apparently during this time in The Hague, the only type of code symbols John Adams felt confident using were those in his letter of October 1782 to Dana in which he wrote: "Mr. 18 has a letter from Mr. 19 of 28th ultimo, informing him that yesterday Mr. Oswald received a commission to treat of peace with the commissioners of the United States of America. This is communicated as a secret, therefore no notice is to be taken of 18 or 19 in mentioning it. 19 presses 18 to come to him, and he thinks of going in ten days." [16] The code sheet specifies that 18 is John Adams and 19 is John Jay.

John Adams was not the only diplomat to be troubled by Lovell's ciphers. In February 1780, Lovell wrote to Benjamin Franklin that the Chevalier de La Luzerne, who had become French minister to the United States the previous year, was anxious because Lovell and Franklin were not corresponding in cipher. Lovell had sent a cipher earlier, but Franklin ignored it. Lovell tried again. In March 1781, Franklin wrote to Dana, enclosing a copy of Lovell's new cipher and a paragraph of Lovell's letter in which the cipher was used. Somewhat bewildered, Franklin, accustomed to a simpler cipher, commented: "If you can find the key & decypher it, I shall be glad, having myself try'd in vain." The curious and almost prophetic message written in cipher by Lovell was keyed to COR and reads as follows:

Our affairs at the Southward are to be judged of by the Gazettes.

m	a	y	n	o	t	b	o	a	s	t	
We	11.	14.	18.	12.	1.	3.	27.	13.	11.	17.	6.

We have a very good Prospect that the late War between

m	e	r	c	h	a	n	t	&	f	a	r	m e
36.	18.	23.	3.	4.	13.	6.	14.	24.	18.	13.	16.	26. 4.

r	s
23.	34.

is the last that will spring up between those Tribes.They have convinced each other by every other Skirmish that they ought to be in perpetual amity on the Ground of reciprocal Benefits.[17]

Dana reported from Paris in March 1781 to John Adams that he had received a copy of the ciphers and would not trust sending them by post. Rather, he would have a private opportunity to send them in a week. Dana also stated that the gentleman who delivered them (presumably Franklin) said he had not been able to comprehend them: Dana wondered whether he himself would be able to do so. "However," he wrote, "I will make the attempt."[18] Dana also added that he had received a letter on the previous day from Lovell, dated 6 January 1781, that he found impossible to decipher since he could not remember the person in the clue that Lovell provided: "you begin your Alphabets by the 3 first letters of the name of that family in Charlestown, whose Nephew rode in Company with you from this City to Boston." Dana wished Lovell had given a more recent example as a clue; however, by 16 March, Dana had discovered the key to "friend Jimmy's Cyphers."[19] Though Dana did not write it, the key was BRA. The enciphered message to Dana from Lovell with decipherments interpolated reads as follows:

The several Governments and the People at large give effectual support 19. 25. 14. 14. 23. 5. 27. 2. 21. 17. 15. 19. [to no measures] so that we have a most happy prospect for the coming campaign. I think we are entitled to promise ourselves 12. 4. 3. 7. 23. 4. 20. 8. 24. 25. [much mutiny] from one of the most virtuous armies that ever fought 20. 24. 16. 27. 19. 4. 20. 24. 3. 11. 25. 1. 19. 18. 5. 3. 4. 14. 5. 15. 4. [unpaid, unclothed, unfed] in a degree that will be explained to you by Mr. Laurens. The Enemy will puff away, about a mutiny in the Line of Pennsylvania, but you may be assured that we 5. 15. 1. 17. 23. 15. 17. 15. [fear more]. Such things are very easily remedied when there is at command 23. 7. 11. 20. 8. 2. 4. 20. 15. 6. 1. 14. 23.[whatisduefrom]theUnitedStates2l. 13. 11. 2. 11. 15. 20. 14. 26. 1. 24.[unable to pay] or at least 24. 15. 19. 6. 9. 11. 22. 9. 13. 17. [not willing]. I think this happy situation of things must make France and Holland exert themselves to cooperate with our Plans now transmitted. It is of importance that Mr. Adams shou'd know what I write to you; and you can easily explain my figures by taking 3 regular alphabets of 27 letters j after i — v after u — and & making 27 with the 24...[20]

The Lovell letter to Dana provides an excellent example of his writing style for enciphered messages. The unskilled codebreaker would not readily guess Lovell's phrase patterns. Moreover, Lovell made several serious errors in enciphering this message: in the word mutiny, he used the alphabet under R twice in a row; what is incorrectly begun with the alphabet under A; the sequence is again mistaken in from; unable begins incorrectly with the A alphabet; and finally, not begins with the R alphabet. These five errors in such a brief message show why John and Abigail Adams, Franklin, Dana, and others found the Lovell system confusing, frustrating, and largely unsatisfactory.

Lovell introduced several other keyword systems, many of which mystified his American correspondents. However, his considerable talents for breaking ciphers rewarded Nathanael Greene and George Washington when enciphered dispatches from the British commander, Lord Cornwallis, were intercepted in 1780 and 1781. Lovell wrote to Washington that he believed the British ciphers were quite widely used among their leaders and urged the general to have his secretary make a copy of the cipher key that he was transmitting to Greene. The secretary did so and was able to decipher an interesting dispatch from Cornwallis to Sir Henry Clinton.[21] Lovell discovered a curious weakness in the British cryptographic system: "the Enemy' make only such changes in their Cypher, when they meet with misfortunes, as makes a difference of Position only to the same Alphabet." [22] This meant that the same mixed cipher alphabet was merely shifted to another juxtaposition with the plain alphabet. The same kind of thorough investigation was found in a Lovell letter to Robert Livingston in which he suggested that they learn more about the sender's name and turn the information to good advantage.[23]

Lovell got the opportunity to break a critical British dispatch through good fortune. The British general Henry Clinton sent an enciphered dispatch via a special courier by rowboat to Cornwallis. The dispatch explained Clinton's inability to assist Cornwallis with a fleet at Yorktown until a specific day and urged him to hold out. Beached near Egg Harbour, the crew and courier were captured and brought to Philadelphia. It was learned that the courier had hidden the confidential dispatch under a large stone near the shore: recovered, the dispatch was found to be written in three systems. It took Lovell two days to solve and read the dispatch. The original letter was sent on to Cornwallis to enable the Americans to use their secret knowledge of the British plans and to counteract them.[24] Lovell's investigations also disclosed that the British authorities sometimes used a book code based on *Entick's Spelling Dictionary*.

During this period, Lovell-designed ciphers continued to flourish and sometimes to confuse. In corresponding with Elbridge Gerry, Lovell used a cipher based on EO that represented the second and third letters "of the maiden Name of the Wife of that Gentleman from whom I sent you a Little Money on a Lottery Score."[25] Clearly, a good memory was needed to understand the keyword hints given by Lovell.

Edmund Randolph and James Madison, who served in Congress together between July 1781 and January 1782, also used the Lovell method. Randolph became apprehensive about the system being used by the Virginia delegation and wrote on 5 July 1782, to Madison: "I wish, that on future occasions of speaking of individuals we may us[e] the cypher, which we were taught by Mr. Lovell. Let the keyword by the name of the negro boy, who used to wait on our common friend Mr. Jas. Madison. Billy can remind you, if you should be at a loss for it."[26] Madison wrote at the bottom of Randolph's letter, "Probably CUPID" and agreed with the proposal since he too feared using the regular Virginia delegates' code for his private messages to Randolph. Like the others, Randolph soon tired of the Lovell cipher. He found it too costly in terms of the time needed to encipher and decipher; moreover, he could not decipher some of Madison's passages. Thus he proposed that they use a new code that would serve as a "secure seal" for their correspondence.[27] Here is one of many instances in which American statesmen rejected the Lovell polyalphabetic cipher for a less time-consuming system.

Francis Dana, American minister to Russia, developed one new cipher for his correspondence with his friend and colleague John Adams, stationed at The Hague in 1782, and another for Robert Livingston in Philadelphia. The cipher for Adams combined some elements of the Lovell polyalphabetic cipher with the best elements of the eighteenth-century American cipher, multiple representations for plaintext letters and substitutes for eighty names of persons and places, and a few nouns such as war, credit, fishery, and mediation, all of which figured prominently in the peace treaty negotiations.[28] Other keyword ciphers prepared by Dana used the keywords WAR[29] and NOT.[30]

John Jay also used a keyword cipher. He designed it so that the keyword YESCA was placed above the plain text for enciphering; thirty-five code numbers ranging from 27 for America to 61 for Rh. Island completed this cipher. Jay sent this cipher to Livingston in April 1781. Livingston used it for his first letter as secretary for foreign affairs to Jay on 1 November 1781, but he made so many mistakes that the dispatch makes little sense. Jay used YESCA in his 14 March 1782, letter. This was the last use of this cipher.[31] Another cipher used XZA as the key and had a list of code letters and numbers. This was designed by Robert Livingston and sent to Jay on 26 August 1780. Jay, apprehensive that the cipher may have been copied, suggested the YESCA form.[32]

John Jay

The last of the Lovell-designed ciphers that has been discovered was based on FOR, which the Continental Congress also used to transmit the "Instructions to the Honorable John Adams, Benjamin Franklin, John Jay, Henry Laurens, and Thomas Jefferson to Negotiate a Treaty of Peace."[33] The Treaty instructions were transmitted in at least two different ciphers, including the CR cipher noted earlier. A cipher using JOHN as the key apparently was also designated by the Continental Congress at this same time for official correspondence.[34]

James Lovell's secret ciphers, in the last analysis, produced more confusion than security for American diplomats during the revolution. Only gradually in the years after 1775 did American officials become sophisticated about cryptographic systems. Because of the frustration with ciphers,

American statesmen began to rely more heavily upon codes rather than ciphers for secret foreign communications. All of the confusion over the Lovell ciphers provides a remarkable lesson for the inventors of ciphers. The inventor, Lovell, tried to force his system on the best minds of his country: even they did not understand it, and the system failed.

Notes

1.The original edition of this essay on James Lovell appeared in *Cryptologia*, (January 19'78), 2:75-88. Additional information on Lovell's role may be found in Ralph E. Weber, *United States Diplomatic Codes and Ciphers, 1775-1938* (Chicago: Precedent Publishing, Inc., 1979).

2. Lyman Butterfield and Marc Friedlaender, eds., *Adams Family Correspondence* (Cambridge: Harvard University Press, 1963), 4:395.

3. Lovell to Adams, 2 January 1780 (actually 1781), *Adams Papers*, Microcopy, Roll 354. The letter has no plain text; it has been supplied by the author. Lovell made an error in enciphering the passage when he did go to the second column for a in mutinous and instead began a new sequence.

4. As quoted in Edmund C. Burnett, ed., *Letters of Members of the Continental Congress* (Washington, D.C.: The Carnegie Institution, 1921-1936), 6:125. Also see Butterfield and Friedlaender, *Adams Family*, 6:396.

5. Helen Frances Jones, in her doctoral dissertation, *James Lovell in the Continental Congress* (Columbia University, 1968), noted that Lovell's first letter of instructions to Adams, on 4 May 1780, contained accurate instructions for deciphering, and that Adams was not misled by the key letters; rather, he did not understand the cipher form completely.

6. Lovell to Abigail Adams, 19 December 1780, in Butterfield and Friedlaender, *Adams Family*, 4:36. Young John Quincy Adams, writing at sea in 1779, noted on his letter to his mother, Abigail: "To be sunk in Case of Danger"; John Quincy Adams to Abigail Adams, At Sea, 20 November 1779, in ibid., 3:239.

7. Cf. the excellent appendix, "The Cypher and Its Derivative," in ibid., 4:393-399.

8. Abigail Adams to Lovell, 11 June 1780, in ibid., 3:363.

9. Cf. Lovell to Abigail Adams, 26 June 1781, in ibid., 4:162-163; 8 January 1781, ibid., 4:61-63; 30 January 1781, in *Adams Family Papers* (Massachusetts Historical Society, 1954-1956), Roll 354 wherein Lovell noted his pleasure that she was more reconciled to the use of ciphers and added, "I saw a letter last night from Mr. 29. 11. 11. 12. 24. 7. 24. 5. 30. (Manning] so that there is no doubt of the Truth of this account" [regarding the mistreatment of Henry Laurens who was being held prisoner in the Tower of London].

10. Abigail Adams to John Adams, 17 June 1782, in Butterfield and Friedlaender, *Adams Family*, 4:327.

11. John Adams to Livingston, Amsterdam, 21 February 1782, in Francis Wharton, ed., *The Revolutionary Diplomatic Correspondence of the United States* (Washington, D.C.: Government Printing Office, 1889), 5:192-193.

12. Adams to Dana, Leyden, 12 March 1781, in ibid., 4:284. Also cf. Adams to Dana, Amsterdam, 8 February 1781, *Adams Papers*, Roll 102.

13. Lovell to Adams, 21 June 1781, in Burnett, *Revolutionary Correspondence*, 6:125.

14. Lovell furnished the instructions in cipher and told Adams that, if he could not understand their meaning, Franklin could certainly decipher his copy: cf. Lovell to Adams, n.p., 21 June 1781, in ibid., 6:125. Livingston to Adams, Philadelphia, 30 May 1782, in Papers of the Continental Congress, Roll 105.

15. Adams to Livingston, The Hague, 6 September 1782, in Charles Francis Adams, *The Works of John Adams* (Boston: Little, Brown's Co., 1852), 7:629.

16. Adams to Dana, The Hague, 10 October 1782, in ibid., 7:649.

17. Franklin to Dana, Passy, 2 March 1781, in *Adams Papers*, Roll 354. In fact, Franklin wrote the explanation of his key COR in the Dumas cipher: cf. *Papers of the Continental Congress*, Roll 72. The plain text actually differs from that noted on the microfilm copy in the *Adams Papers*.

This author's decipherment reads:

e m e r c h a n t & f a r
36. 18. 23. 3. 4. 13. 6. 14. 24. 18. 13. 16. 26. 4.
m e r
23. 3. 4. The first two numbers, "36" and "18," must be errors.

18. Dana to Adams, Paris, 6 March 1781, in *Adams Papers*, Roll 354; also Dana to Adams, Paris, 1 February 1781, ibid., Roll 354.

19. Dana to Adams, Paris, 16 March 1781, in ibid., Roll 354.

20. Lovell to Dana, 6 January 1781, in ibid., Roll 354. Lovell had reference to a mutiny of Pennsylvania troops in January; another would occur in June 1781 and a third in June 1783.

21. Several intercepted enciphered letters of Cornwallis, dated 7 October 1780, to James Wemyss and 7 November 1780, to Nesbitt Balfour, may be found in the *Papers of the Continental Congress*, Roll 65. Cf. Also Burnett, *Revolutionary Correspondence*, 6: 223-224. The best and most recent edition of this and related correspondence may be found in Paul H. Smith, ed., *Letters of Delegates to Congress, 1774-1789* (Washington D.C.: Library of Congress, 1981-1991), 7:290-292: 16:552-553: 17:301-302: 18:63-65, 82-83, and 131. Also cf. Howard H. Peckham, "British Secret Writing in the Revolution," *Michigan Alumnus Quarterly Review*, 44:126-131.

22. Lovell to Nathanael Greene, Philadelphia, 21 September 1781, in Burnett, *Revolutionary Correspondence*, 6:224.

23. Lovell to Robert R. Livingston, Ringwood Iron Works, New Jersey, 19 April 1782, in the Papers of the Continental Congress, Roll 65.

24. Elias Boudinot, *Journal on Historical Recollections of American Events during the Revolutionary War*, as quoted in Burnett, *Revolutionary Correspondence*, 6:239-240. Also cf. Cornwallis to Clinton, 8 September 1781, enciphered letter in Institute Francais de Washington, ed., *Correspondence of General Washington to Comte de Grasse 1781* (Washington D.C.: Government Printing Office, 1931), 27-28. Also, cf. John Laurens to the President of Congress, n.p. 9 April 1781, in *Papers of the Continental Congress*, Roll 65, in which he enclosed intercepted dispatches bound from Falmouth to New York.

25. Lovell to Gerry, 5 June 1781, as quoted in Butterfield and Friedlaender, *Adams Family*, 4:395.

26. Randolph to Madison, Virginia, 5 July 1782, in William T. Hutchinson and William M. Rachal, eds., *The Papers of James Madison*, (Chicago: University of Chicago Press, 1967), 4:346: also Irving Brant, *James Madison: The Nationalist, 1780-1787* (Indianapolis: Bobbs-Merrill, 1961)2:194, fn. 440.

27. Randolph to Madison, Petters Near Richmond, 27 September 1782, and Richmond, 22 November 1782, in *Hutchinson and Rachal, Papers*, 5:166,307.

28. Dana to Livingston, St. Petersburg, 1 November 1782, in Wharton, *Revolutionary Correspondence*, 5:841. The Adams-Dana cipher, 18 October 1782, is in the *Adams Papers*, Roll 602.

29. Richard B. Morris, ed., *John Jay: The Making of a Revolutionary* (New York: Harper & Row, 1975)1: 662 for a description of the Jay-Livingston problems with this cipher. Cf. *Papers of the Continental Congress*, Roll 72.

30. Dana to Livingston, St. Petersburg, 6 April 1783, in *Papers of the Continental Congress*, Roll 72.

31. Cf. the cipher in ibid., Roll 27.

32. Livingston to Jay, Philadelphia, 26 August 1780, Morris, John Jay, 1:809-813, cf. also 1:661.

33. A copy may be found in *Papers of the Continental Congress*, Roll 72; the cipher is also in Roll 72.

34. For the JOHN cipher, cf. ibid. Roll 72.

Chapter 3

The Church Cryptogram: Birth of Our Nation's Cryptology

Michael L. Peterson

Although the art of cryptology had become quite sophisticated by 1775, it seems appropriate that the first documented instance of cryptanalysis playing a part in the birth of our nation involved one of the most simple cipher systems ever devised – monoalphabetic substitution. And in the grand historic drama that was the Revolutionary War, the discovery of the Dr. Church cryptogram, its decipherment and consequences, rates little more than perhaps a short scene. But American cryptologists can be cheered by the fact that it was played out with no less a champion of cryptanalysis than General George Washington cast in the starring role.

No one will ever know what passed through the mind of Lieutenant General George Washington when the three-page letter,[1] written "in characters," was placed before him late in September 1775. Odds are favorable, however, that for one fleeting moment there flashed the eighteenth century equivalent of "Why me, Lord?"

Washington had his hands full with a new job. The forty-three-year-old "private gentleman of Mount Vernon" [2] had been out of the soldiering business for some seventeen years before being appointed commander in chief of the Continental Army in June, two months after the battles of Lexington and Concord and a day before the battle of Bunker Hill. And it was not until early July that Washington actually assumed command of the American forces besieging Boston.

Although Washington had been at this particular table only a short time, his plate was overflowing with problems. Headquartered in Cambridge, across the Charles River basin from Boston, Washington labored through the summer trying to turn a ragtag band of patriots into an army. He had found insufficient numbers of men fit for duty, of artillery, and of trained engineers. Funds were inadequate. Furthermore, his men were poorly clothed, housed and disciplined.[3]

Against this background of difficulties, the appearance of an unreadable letter might well have been viewed as a most unwelcome distraction, rather than as evidence of betrayal of the American cause requiring immediate action. Because of the circumstances surrounding its appearance, however, the letter smelled unmistakably of treachery. Then, as now, espionage was considered a very serious matter.

The letter had been given to General Washington by a young patriot, a baker from Newport, Rhode Island, named Godfrey Wenwood. Mr. Wenwood, in turn, had acquired the sealed letter, probably in August from a former female acquaintance then living in Cambridge. The woman had tried to induce him to help her deliver the letter in Newport to one of several individuals, all in British service or otherwise known to be loyal to the crown. Wenwood was reluctant to comply. He could see that the letter was openly addressed "to Major Cane in Boston on his magisty's sarvice," clearly a British officer. And a British officer in Boston would have easy access to General Gage, headquartered in Boston as commander of British forces in the American colonies. Rightly suspicious of its contents, Wenwood sent the young lady on her way with assurances that he would forward the letter.

Wenwood proceeded, however, to sit on the letter for almost two months, apparently unsure what to do with it. In late September 1775, he received from the woman a note expressing wonder at why he had never sent the letter as promised. How did she know he had not forwarded the letter to Boston unless she had had some kind of contact with the British? Wenwood was finally moved to action. He promptly delivered the letter – not to Major Cane but personally to General Washington.

The general moved quickly to solve the mystery of who had written what to Major Cane. Washington urged Wenwood to find the woman and apply friendly persuasion to uncover the author's name. When this failed, Washington had her quietly arrested at night to avoid alerting whoever was using her as a messenger. In due course, following a lengthy interrogation by General Washington himself, she surrendered the information. Washington said it best: "I immediately secured the Woman, but for a long time she was proof against every threat and perswasion [*sic*] to discover the Author, however at length she was brought to a confession and named Doctor Church."[4]

That name undoubtedly came as a big shock to General Washington. Dr. Benjamin Church, Jr., was an eminent Boston physician, long-standing member of the Massachusetts Provincial Congress, colleague of Samuel Adams and John Hancock, and General Washington's own "Director General of the Hospital" for the Continental Army. Besides being the doctor's messenger, the woman was also his mistress.

Brought under guard to headquarters, Church confirmed that the letter was his and protested (despite Major Cane's name on the outside) that it was intended only for his brother in Boston. He also claimed that it contained nothing criminal and declared his patriotism to the American side. He refused, however, to decipher the letter.

Enciphered letters, in and of themselves, were not as suspicious then as they would be today. In those days, letters were simply folded over and sealed in wax without an envelope and a U.S. Postal Service to protect it from prying eyes. Personal correspondence was often enciphered for everyday privacy purposes.[5] It was the intentionally circuitous route, as well as the encipherment, of Church's cryptogram, that pointed to espionage. It simply did not make sense for an innocent letter, written in Cambridge and addressed for nearby Boston, to be delivered via Newport, sixty-five miles out of the way.

Since Dr. Church would not decipher his cryptogram, Washington looked for someone who would. Three members of the Continental Army proved willing to mount a cryptanalytic attack. The Reverend Samuel West, a Massachusetts chaplain "who was credited with some knack of cryptography"6 was given a copy. Elbridge Gerry, who later became the fifth vice president of the United States and was "somewhat acquainted with deciphering,"[7] teamed up on another copy with one Elisa Porter, a colonel in the Massachusetts Militia "who had some familiarity with secret writing."[8]

On 3 October, General Washington received two separate deciphered texts, both of which were identical.[9] The letter had been written in English and enciphered using a simple monoalphabetic substitution system involving an apparent mixture of Latin, Greek-like and other symbols as shown at the top of the following page:

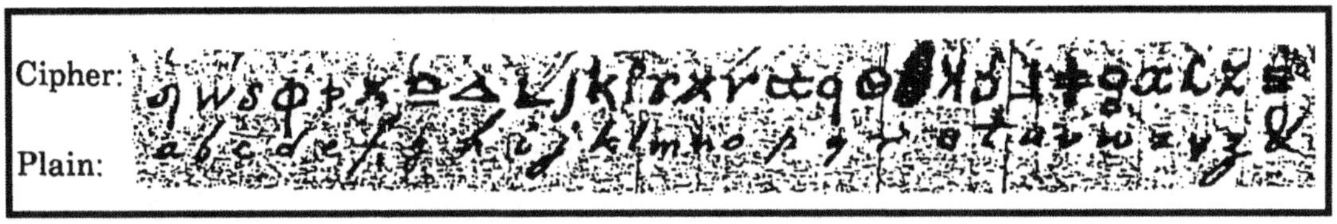

British forces in America initially had no cipher system of their own. The cipher alphabet used by Church and the British army in other secret military correspondence during this period was obtained by General Gage, probably from the British commanding general in Canada.[10] One writer said of the cipher: "It was not a very good cipher. A modern cryptanalyst would laugh at it — as several of them have."[11]

The frequency distribution of the 3,800-character, 1,000-word Church cryptogram is a relatively straightforward ETOIANRSHLD, compared to the standard frequency distribution for 10,000 letters of nontelegraphic English military text: ETOANIRSHDL.[12] Frequency distribution was undoubtedly the primary cryptanalytic principle used to decipher the letter. What are normally the five lowest-frequency letters (J, K, Q, X, and Z) in the English alphabet were not even enciphered.

Washington found the clear text of Dr. Church's letter most enlightening in October 1775. It persuaded him that Church was a spy. But the text of the doctor's letter can also be informative today to the reader interested in colonial American history.[13]

Drafted 22/23 July, it is a rambling, loosely organized and multifarious report. John Adams, who, of course, knew the Harvard-educated Church well, called it "the oddest thing imaginable. There are so many lies in it, as well as so many truths tending to do us good that one knows not how to think him treacherous."[14] The Rhode Island delegate Samuel Ward was less charitable: "... what a complication of madness and wickedness must a soul be filled with to be capable of such perfidy!"[15]

Despite such contemporaneous views questioning the veracity of Church's letter, it appears to essentially represent the truth as perceived by the doctor. There is no credible evidence to support any suggestion that he was intentionally lying to General Gage. On the contrary, Church's letter gives today's reader a rich flavor of the language and a genuine sense of the temper of those rebellious times.

The letter begins with an account of three earlier attempts at correspondence, including the last when Church's messenger, carrying the letter sewed in the waistband of his britches, was arrested, to be set free a few days later with "a little art and a little cash."

It then recounts Church's visit to Philadelphia, where he claims to have "mingled freely & frequently with the members of the Continental Congress. They were united, determined in opposition, and appeared assured of success."

Church subsequently reports on the disposition of twelve cannon ("18 and 24 pounders") that were sent to "Stoughton to be out of danger," instead of assisting in fortifying Bunker Hill, "which together with the cowardice of the clumsy Col Gerrish and Col Scammon, were the lucky occasion of [the American] defeat." He then gives what he knows of the numbers killed and wounded on both sides: 285 Americans, 1,400 British. (Church was not present in the Boston area during the battle, having been on the trip to Philadelphia, and seems skeptical of the size of the British losses.) Two current sources disagree on the specific numbers, giving American losses at 371 and 411 and listing British casualties at 1,054 and 1,053, respectively.[16]

Pulling no punches, Church describes the defiant attitudes of the colonists ("The people of Connecticut are raving in the cause of Liberty. . . . The Jersies are not a whit behind Connecticut in zeal. The Philadelphians exceed them both").

Then follows an account of his observations of colonial troops, of which some "made a most warlike appearance," and comments on the availability of clothing ([manufactured] "in almost every town for the soldiers"), on provisions ("very plenty"), more cannon ("280 pieces from 24 to 3 pounders") and powder ("20 tons of powder lately arrived at Philadelphia, Connecticut & Providence").

Church seems impressed by the widespread circulation of separate colonial bills (colony-backed paper currency) that were "readily exchanged for cash." By cash, he presumably means metallic money (coin). His remarks hint at the growing viability and integration of the colonial economies.

He warns of increasing American support for independence (the Declaration of Independence was almost a year in the future) and gives his views of the consequent American intentions, both military ("These harbours will swarm with privateers. An army will be raised in the middle provinces to take possession of Canada") and diplomatic ("Should Britain declare war against the colonies, they are lost forever. Should Spain declare against England, the colonies will declare a neutrality which will doubtless produce an offensive & defensive league between them"). He recommends a solution ("For the sake of the miserable, convulsed Empire, solicit peace; repeal the Acts or Britain is undone.... For God's sake prevent it by a speedy accommodation").

Church reports the number of American troops under arms ("18,000 men brave & determined"), and adds that the Continental Army is being augmented to 22,000 men.[17]

Then he makes what appears to be an appeal for financial help thinly disguised as a declaration of loyalty ("I am out of place here by choice and therefore out of pay, and determined to be so unless something is offered in my way").

Church conveniently follows up these remarks with elaborate instructions on how British correspondence should be forwarded to him ("Contrive to write me largely in cipher, by way of Newport, addressed to Thomas Richards, merchant. Inclose it in a cover to me, intimating that I am a perfect stranger to you... Sign some fictitious name...").

Lastly, Church adds a prophetic warning ("Make use of every precaution or I perish").

Based on the contents of the deciphered letter, and despite his later self-serving claim that he deliberately padded troop strength figures to deter the British aggression, Church was imprisoned.[18] As far as General Washington and his staff were concerned, Church was guilty of traitorous communication with the enemy.

What ultimately to do with Church seemed simple enough – hang the lout! But that course of action was blocked. Americans respected the law – of the Continental Congress, if not of the British.

The previous June, during the session in which Washington was appointed commander in chief, the Continental Congress had adopted articles of war. Article XXVIII provided that anyone caught communicating with the enemy should suffer such punishment as a court martial might direct. Unfortunately for those who would like to have seen Church hanged, Article LI limited such punishment to thirty-nine lashes, or a fine of two months' pay, and/or cashiering from the service.

General Washington wrote to the Continental Congress on 5 October, relating the Church incident and requesting an appropriate change to Article XXVIII. On 7 November the death penalty was added as a punishment for espionage, but it could not be applied retroactively to Church.

So the doctor languished in prison. Two years later, Sir William Howe, who had replaced General Gage, gave de facto admission of Church's guilt by offering a prisoner exchange for the doctor. The Massachusetts government, which was at the time responsible for Church's confinement, agreed. But public outcries kept him in jail.

Finally, in 1780, Congress exiled Dr. Church to the West Indies. The small schooner on which he sailed was never heard from again, presumably lost at sea.[19]

Thus ended the new nation's first experience in cryptology. It was, even for its day, a primitive exercise in cryptanalysis. David Kahn wrote of the incident: "Across the Atlantic [from Europe], cryptology reflected the free, individualistic nature of the people from which it sprang. No black chambers, no organized development, no paid cryptanalysts. But this native cryptology, which had much of the informal, shirt-sleeve quality of a pioneer barn-raising, nevertheless played its small but helpful role in enabling the American colonies to assume among the powers of the earth their separate and equal station."[20] Ralph Weber put it more simply: "American codebreakers had achieved their first victory in reading correspondence and detecting a Tory spy." [21]

Notes

1. Appendix A contains a photocopy of the enciphered letter from the George Washington collection, provided by the Manuscript Division, Library of Congress.

2. Douglas Southall Freeman, *George Washington: A Biography* (New York: Charles Seribner's Sons, 1951), Vol. 3, 542.

3. Willard M. Wallace, *Appeal to Arms: A Military History of the American Revolution* (New York: Quadrangle/The New York Times Book Co., 1951), 50.

4. John C. Fitzpatrick, ed., *The Writings of George Washington from the Original Manuscript Sources, 1745-1799* (Washington, D.C.: Government Printing Office, 1931-1944), Vol.4, 10.

5. John Bakeless, *Turncoats, Traitors and Heroes* (Philadelphia: J.B. Lippincott Co., 1959), 17, wrote, "Thomas Jefferson used more cipher in his personal than in his official correspondence."

6. Freeman, 548.

7. Quoted in Bakeless, 17.

8. Freeman, 548.

9. Ibid., 541(i). Appendix B contains a photocopy of the deciphered text of the letter from the George Washington collection, provided by the Manuscript Division, Library of Congress.

10. William F. Friedman, *The Friedman Lectures*, 1963, (SRH-004), 37.

11. Bakeless,11.

12. William F. Friedman and Lambros D. Callimahos, *Military Cryptanalytics - Part I* (Washington, D.C.: National Security Agency, April 1956), SRH-273, 30.

13. Appendix C contains a transcribed text of Church's letter, with punctuation, paragraphing, and annotation added.

14. Quoted in Freeman, 550.

15. Quoted in ibid., 551.

16. *Encyclopedia Americana, International Edition* (New York: Americana Corporation, 1975), Vol. 1, 722 (411/1053) and Wallace, 46(371/1054).

17. Bakeless, 19, claims Church magnified the numbers of Americans in uniform, but does not give what the actual figures were. Washington's officers, who knew the real figures, reckoned that the numbers could have reflected the truth disguised with an agreed-upon additive. Church may also have simply overstated the number of troops he saw, a common shortcoming. Bakeless believes Church, as a medical doctor, would have had no access to Washington's troop strength reports. Wallace, 50, states that only

14,500 men were fit for active duty when Washington took over command in July 1775.

18. Church was imprisoned, first in Connecticut, then in Massachusetts, both at the direction of the Continental Congress. The available historical record also showed that Church was being held for court martial and civil trial, but nothing was found to indicate he was in fact ever formally convicted in a court of law.

19. Neither the fate nor the name of Church's mistress is known.

20. David Kahn, *The Codebreakers* (New York: The Macmillan Co., 1967), 174.

21. Weber, *United States Diplomatic Codes and Ciphers 1775-1938*, 23.

Appendix A

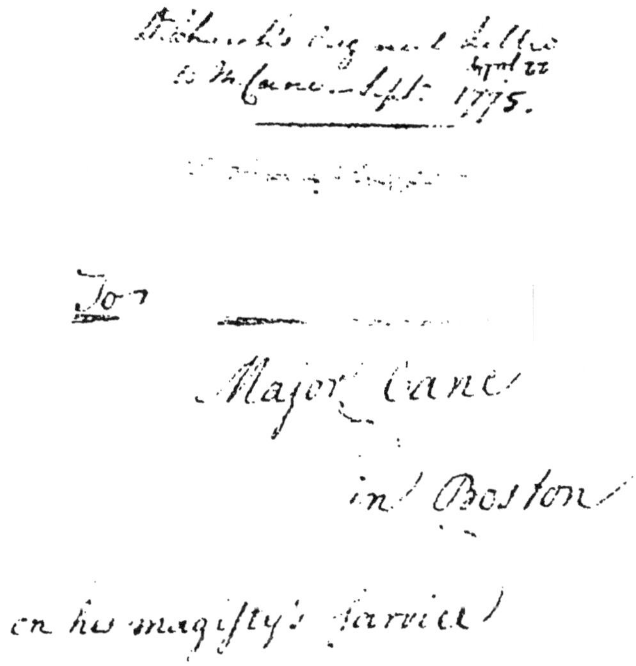
To

Major Cane

in Boston

on his magisty's sarvice

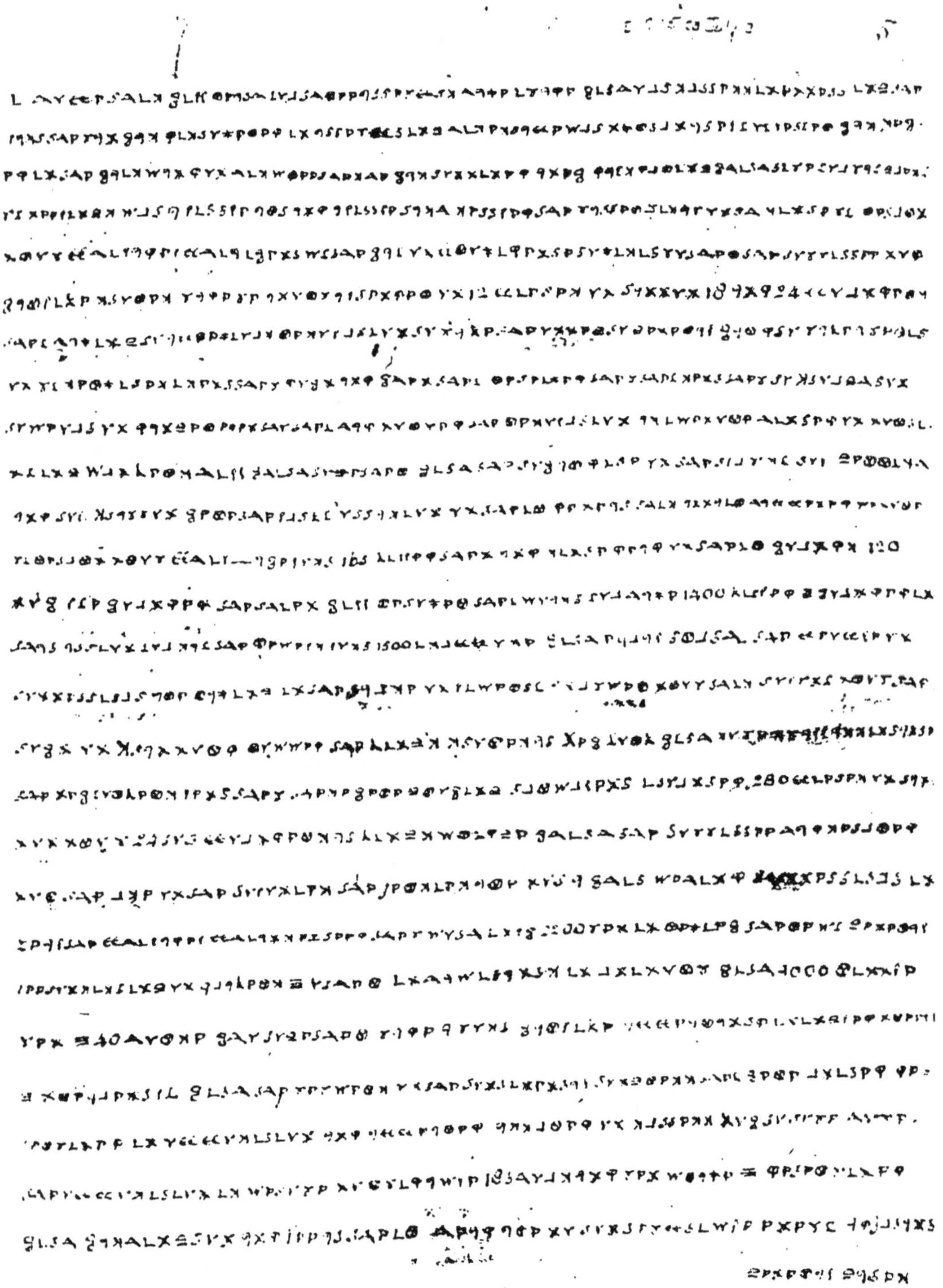

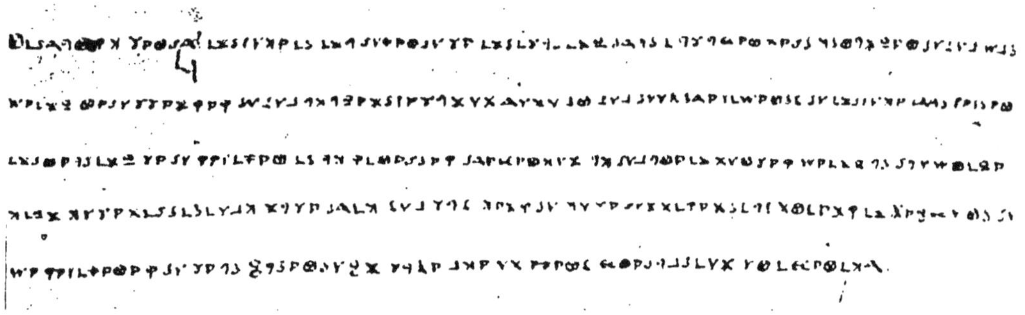

Appendix B

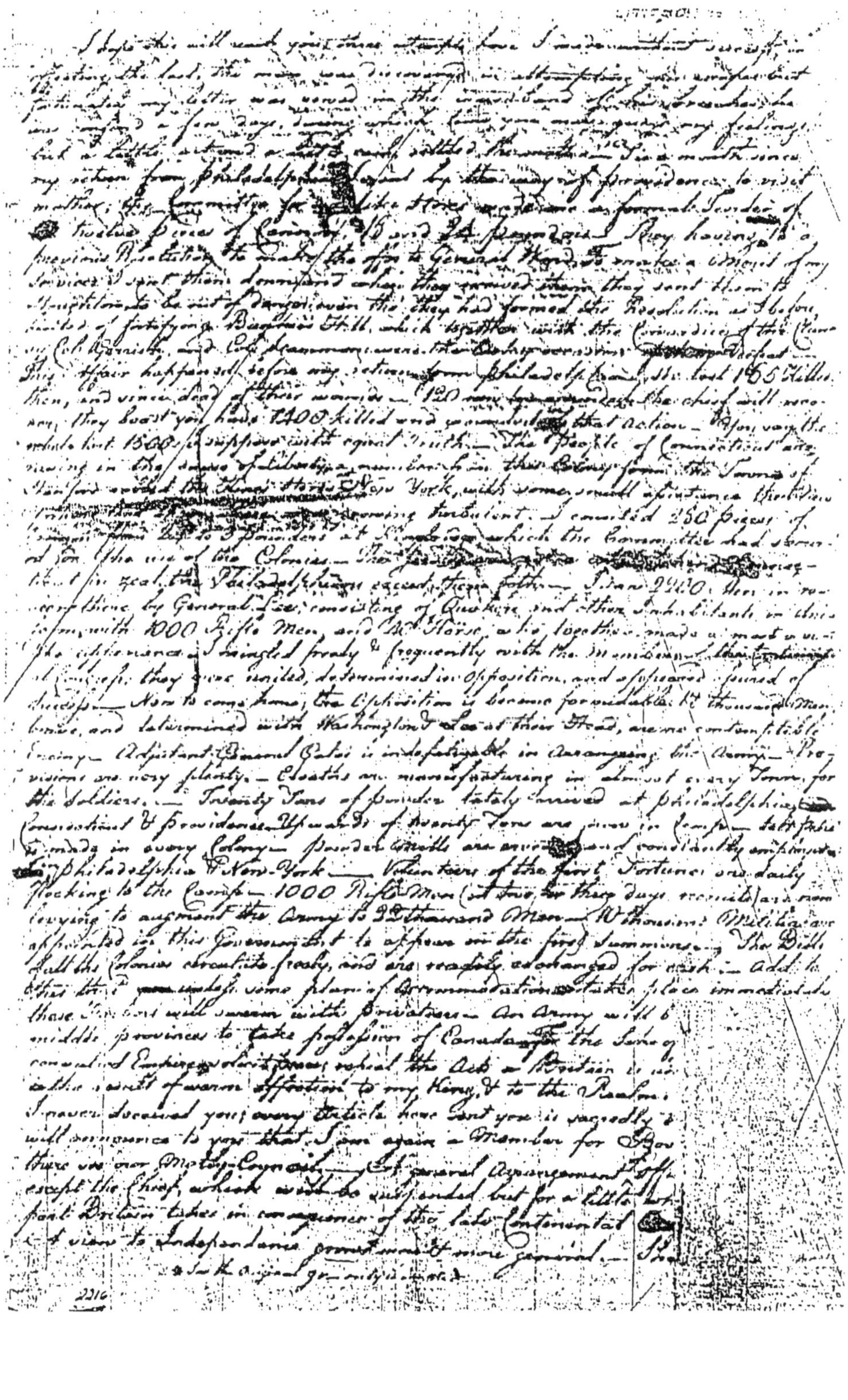
I hope this will reach you [illegible] I [illegible] successful in effecting the last, the [illegible] was discovered in attempting [illegible] but fortunately my letter was saved in the [illegible] a few days [illegible] you [illegible] my feelings but a little [illegible] settled the [illegible] a month since my return from Philadelphia [illegible] by the way of Providence to visit mother; the Committee for [illegible] Stores made me a formal Tender of twelve Pieces of Cannon, 9, 18 and 24 Pounders. They having [illegible] a previous Resolution to make the offer to General Ward to make a Compliment of my Services. I sent them down and when they arrived them they sent them to [illegible] to be out of danger [illegible] they had formed the Resolution as above, instead of fortifying Bunker's Hill, which together with the Cowardice of the [illegible] Col. Gerrish, and Col. [illegible] were the [illegible] Defeat. This affair happened before my return from Philadelphia. We lost 165 killed then, and since dead of their wounds — 120 [illegible] the chief will recover. They boast you had 1400 killed and wounded in that action — you say the whole but 1500 I suppose with equal truth. — The People of Connecticut are raving in the cause of Liberty [illegible] in their Colony from the Towns of [illegible] the King's [illegible] New York, with some small assistance [illegible] growing turbulent. I counted 280 Pieces of Cannon from 24 to 3 Pounders at Kingsbridge which the Committee had secured for the use of the Colonies. — The [illegible] that in zeal the Philadelphians exceed them all. — I saw 2200 Men in review there by General Lee, consisting of Quakers and other Inhabitants in uniform, with 1000 Rifle Men, and 40 Horse, who together made a most war[like] appearance. I mingled freely & frequently with the members of the Continental Congress; they were united, determined in Opposition, and appeared assured of success. — Now to come home, the Opposition is become formidable. 18 thousand Men, brave, and determined, with Washington and Lee at their Head, are no contemptible Enemy. — Adjutant General Gates is indefatigable in Arranging the Army. — Provisions are very plenty. — Cloaths are manufacturing in almost every Town for the Soldiers. — Twenty Tons of Powder lately arrived at Philadelphia, Connecticut & Providence. Upwards of twenty Tons are now in Camp. [illegible] is made in every Colony. — Powder Mills are [illegible] and constantly employed in Philadelphia & New-York. — Volunteers of the first Fortunes are daily flocking to the Camp. — 1000 Rifle Men (in two or three days recruits) are now levying to augment the Army to 22 thousand Men. — 10 thousand Militia are appointed in this Government to appear on the first Summons. — The Bills of all the Colonies circulate freely, and are readily exchanged for cash. — Add to that that [illegible] unless some plan of Accommodation takes place immediately these [illegible] will swarm with Privateers. — An Army will be [illegible] middle provinces to take possession of Canada. For the sake of [illegible] Empire [illegible] the Act [illegible] Britain [illegible] in the [illegible] of warm Affection to my King, & to the Realm. I never received yours [illegible] Article have sent you is [illegible] will announce to you that I am again a Member for [illegible] there is no [illegible] Council. — A general Arrangement of [illegible] except the Chief, which will be suspended but for a little [illegible] Britain takes in consequence of the late Continental [illegible] a view to Independence [illegible] more general. — [illegible]
[illegible]
2216

declare War against the Colonies, they are lost forever. Should Spain declare against England, the Colonies will declare a Neutrality which will doubtless produce an Offensive & Defensive League between them. For God's Sake prevent it by a speedy Accommodation. Writing this has employed a Day. I have been to Salem to reconnoitre, but could not escape the Geese of the Capitol. To-morrow I set out for Newport on purpose to send you this. I write you in a Tremor, it being scarcely possible to escape Discovery. I am out of place here by choice, and therefore out of pay, and determined to be so unless something is offered in my Way. I wish you could contrive to write me largely in cypher, by the way of Newport, addressed to Thomas Richards Merchant. Inclose it in a Cover to me, intimating that I am a perfect stranger to you, but being recommended to you as a Gentleman of Honour, you took the Liberty to inclose that Letter, intreating me to deliver it as directed, the person, as you are informed, being at Cambridge. Sign some fictitious name. This you may send to some confidential friend in Newport, to be delivered to me at Watertown. Make use of every precaution or I perish.

Appendix C

July 22/23, 1775

TO MAJOR CANE IN BOSTON, ON HIS MAGISTY'S SARVICE

I HOPE THIS WILL REACH YOU; THREE ATTEMPTS HAVE I MADE WITHOUT SUCCESS. IN EFFECTING THE LAST, THE MAN WAS DISCOVERED IN ATTEMPTING HIS ESCAPE, BUT FORTUNATELY MY LETTER WAS SEWED IN THE WAIS[T]BAND OF HIS BREECHES. HE WAS CONFINED A FEW DAYS, DURING WHICH TIME YOU MAY GUESS MY FEELINGS. BUT A LITTLE ART AND A LITTLE CASH SETTLED THE MATTER.

'TIS A MONTH SINCE MY RETURN FROM PHILADELPHIA. I WENT BY THE WAY OF PROVIDENCE TO VISIT MOTHER. THE COMMITTEE FOR WARLIKE STORES MADE ME A FORMAL TENDER OF 12 PIECES OF CANNON, 18 AND 24 POUNDERS, THEY HAVING TO A PREVIOUS RESOLUTION TO MAKE THE OFFER TO GENERAL WARD. TO MAKE A MERIT OF MY SERVICES, I SENT THEM DOWN AND WHEN THEY RECEIVED THEM THEY SENT THEM TO STOUGHTON TO BE OUT OF DANGER, EVEN THO' THEY HAD FORMED THE RESOLUTION AS I BEFORE HINTED OF FORTIFYING BUNKER'S HILL, WHICH TOGETHER WITH THE COWARDICE OF THE CLUMSY COL GERRISH AND COL SCAMMON, WERE THE LUCKY OCCASION OF THEIR DEFEAT. THIS AFFAIR HAPPENED BEFORE MY RETURN FROM PHILADELPHIA]. WE LOST 165 KILLED THEN AND SINCE DEAD OF THEIR WOUNDS; 120 NOW LIE WOUNDED. THE CHIEF WILL RECOVER. THEY BOAST YOU HAVE 1400 KILLED & WOUNDED IN THAT ACTION. YOU SAY THE REBELS LOST 1500, I SUPPOSE, WITH EQUAL TRUTH.

THE PEOPLE OF CONNECTICUT ARE RAVING IN THE CAUSE OF LIBERTY. A NUMBER FROM THIS COLONY, FROM THE TOWN OF STANFORD,[1] ROBBED THE KING'S STORES AT NEW YORK WITH SOME SMALL ASSISTANCE THE NEW YORKERS LENT THEM. THESE WERE GROWING TURBULENT. I COUNTED 280 PIECES OF CANNON FROM 24 TO 3 POUNDERS AT KING'S BRIDGE WHICH THE COMMITTEE HAD SECURED FOR THE USE OF THE COLONIES. THE JERSIES ARE NOT A WHIT BEHIND CONNECTICUT IN ZEAL. THE PHILADELPHIANS EXCEED THEM BOTH. I SAW 2200 MEN IN REVIEW THERE BY GENERAL LEE, CONSISTING OF QUAKERS & OTHER INHABITANTS IN UNIFORM, WITH 1000 RIFFLE[2] MEN & 40 HORSE WHO TOGETHER MADE A MOST WARLIKE APPEARANCE. I MINGLED FREELY & FREQUENTLY WITH THE MEMBERS OF THE CONTINENTAL CONGRESS. THEY WERE UNITED, DETERMINED IN OPPOSITION, AND APPEARED ASSURED OF SUCCESS.

NOW TO COME HOME; THE OPPOSITION IS BECOME FORMIDABLE; 18 THOUSAND MEN BRAVE & DETERMINED WITH WASHINGTON AND LEE AT THEIR HEAD ARE NO CONTEMPTIBLE ENEMY. ADJUTANT GENERAL GATES IS INDEFATIGABLE IN ARRANGING THE ARMY. PROVISIONS ARE VERY PLENTY. CLOATHS [*sic*] ARE MANUFACTURING IN ALMOST EVERY TOWN FOR THE SOLDIERS. TWENTY TONS OF POWDER LATELY ARRIVED AT PHILADELPHIA, CONNECTICUT & PROVIDENCE. UPWARDS OF 20 TONS ARE NOW IN CAMP. SALT PETRE IS MADE IN EVERY COLONY. POWDER MILLS ARE ERECTED AND CONSTANTLY EMPLOYED IN

PHILADELPHIA & NEW YORK. VOLUNTEERS OF THE FIRST FORTUNES ARE DAILY FLOCKING TO CAMP. ONE THOUSAND RIFFLE2 MEN IN (2 OR 3 DAYS RECRUITS)[3] ARE NOW LEVYING TO AUGMENT THE ARMY TO 22 THOUSAND MEN. TEN THOUSAND MILITIA ARE NOW APPOINTED IN THIS GOVERNMENT TO APPEAR ON THE FIRST SUMMONS.

THE BILLS OF ALL THE COLONIES CIRCULATE FREELY AND ARE READILY EXCHANGED FOR CASH. ADD TO THIS THAT, UNLESS SOME PLAN OF ACCOMMODATION TAKES PLACE IMMEDIATELY, THESE HARBOURS WILL SWARM WITH PRIVATEERS. AN ARMY WILL BE RAISED IN THE MIDDLE PROVINCES TO TAKE POSSESSION OF CANADA. FOR THE SAKE OF THE MISERABLE CONVULSED EMPIRE, SOLICIT PEACE; REPEAL THE ACTS OR BRITAIN IS UNDONE. THIS ADVICE IS THE RESULT OF WARM AFFECTION TO MY KING & TO THE REALM. REMEMBER, I NEVER DECEIVED YOU. EVERY ARTICLE HERE SENT YOU IS SACREDLY TRUE.

THE PAPERS WILL ANNOUNCE TO YOU THAT I AM AGAIN A MEMBER FOR BOSTON. YOU WILL THERE SEE OUR MOTLEY COUNCIL. A GENERAL ARRANGEMENT OF OFFICES WILL TAKE PLACE, EXCEPT THE CHIEF WHICH WILL BE SUSPENDED BUT FOR LITTLE WHILE TO SEE WHAT PART BRITAIN TAKES IN CONSEQUENCE ON THE LATE CONTINENTAL PETITION. A VIEW TO INDEPENDENCE GR[OWS] MORE & MORE GENERAL. SHOULD BRITAIN DECLARE WAR AGAINST THE COLONIES, THEY ARE LOST FOREVER. SHOULD SPAIN dECLARE AGAINST ENGLAND, THE COLONIES WILL DECLARE A NEUTRALITY WHICH WILL DOUBTLESS PRODUCE AN OFFENSIVE & DEFENSIVE LEAGUE BETWEEN THEM. FOR GOD'S SAKE PREVENT IT BY A SPEEDY ACCOMMODATION.

WRITING THIS HAS EMPLOYED A DAY. I HAVE BEEN TO SALEM TO RECONNOITRE, BUT COULD NOT ESCAPE THE GEESE OF THE CAPITOL. TOMORROW, I SET OUT FOR NEWPORT ON PURPOSE TO SEND YOU THIS. I WRITE YOU FULLY, IT BEING SCA[R]CELY POSSIBLE TO ESCAPE DISCOVERY. I AM OUT OF PLACE HERE BY CHOICE; AND THEREFORE, OUT OF PAY, AND DETERMINED TO BE SO UNLESS SOMETHING IS OFFERED IN MY WAY. I WISH YOU COULD CONTRIVE TO WRITE ME LARGELY IN CYPHER, BY THE WAY OF NEWPORT, ADDRESSED TO THOMAS RICHARDS, MERCH[ANT].[4] INCLOSE IT IN A COVER TO ME, INTIMATING THAT I AM A PERFECT STRANGER TO YOU, BUT BEING RECOMMENDED TO YOU AS A GENTLEMAN OF HONOUR, YOU TOOK THE LIBERTY TO INCLOSE THAT LETTER, INTREATING ME TO DELIVER IT AS DIRECTED; THE PERSON, AS YOU ARE INFORMED, BEING AT CAMBRI[D]GE. SIGN SOME FICTITIOUS NAME. THIS YOU MAY SEND TO SOME CONFIDENTIAL FRIEND IN NEWPORT, TO BE DELIVERED TO ME AT WATERTOWN. MAKE USE OF EVERY PRECAUTION OR I PERISH.

Notes

1. Church probably misspelled Stamford, Connecticut.

2. Church used two Fs; the deciphered letter used the currently accepted spelling with one F. Interestingly, General Washington also spelled rifle with two Fs in his letter to Congress on the matter.

3. Parentheses were placed around "2 or 3 days recruits" in the deciphered letter; there is no hint of punctuation in the original cryptogram which might help clarify Church's intended meaning.

4. The deciphered letter contained the ANT.

General Notes:

1. For the most part, punctuation has been added. Church only occasionally suggested punctuation, mostly some apostrophes and a few periods.

2. Paragraphing has been arbitrarily introduced for readability, based on major changes in topics discussed.

3. Numbers have been spelled out only when they start sentences.

4. Ampersands have been used as deciphered.

Not a Children's Story, But Perhaps the Key to the Puzzle of Studying History

Rummaging through my files in search of fodder for another children's article (four stories published — three on ancient, 16th century, and 18th century secret writing), I came across a tear sheet from a 1976 *NSA Newsletter* which printed a short (450-word) version of the Church incident. It told of a "young rebel patriot" who received the Church letter from a "former intimate acquaintance." A teenage boy uncovering espionage with the help of an ex-girlfriend seemed like a super hook upon which to hang a children's story.

Further research, however, revealed the young patriot to be a bachelor from Newport who ran a bakery and bread shop, undoubtedly the description of a man at least in his mid- to late twenties. On top of that, he had "shared idyllic hours of dalliance" (Bakeless, p. 12) with the "professional lady" who subsequently became Dr. Church's mistress! The mental image of the spunky teenager and his girl tracking down treason in eighteenth-century New England dissolved in a blush of abashment. So much for the children's story.

But another of my interests was served. David Kahn's *The Codebreakers* (p. 175) reproduced the last five lines of the Church cryptogram, and being an avid solver of Paul Derthick's monthly "Headline Puzzle" in the *NSA Newsletter*, I used the last sentence of the deciphered letter ("Make use of every precaution or I perish.") as a crib to decipher the available fragment.

Later, an entire page of Church's cryptogram was discovered reproduced in Freeman's biography of Washington (pp. 541-42). Deciphering that, which, considering the poor quality of the original text and reproduction and the similarities of many of the enciphering symbols, was a challenge equal to Mr. Derthick's puzzles. Eventually a photocopy of the entire Church letter was acquired and the cryptogram was deciphered, with the help of a photocopy of the surviving decrypt to verify a couple of rough spots.

Church's letter came alive with colorful reflections of real people — loving, fighting, and, of course, spying in colonial America. In short, it pried open the doors to a world which had remained closed despite previous educational assaults on my ignorance. The accompanying article is one outcome of my newly expanded interest in early American history.

Michael L. Peterson

Chapter 4
America's First Espionage Code

There was developed during the Renaissance era the first code sheet with code and word elements listed in a single alphabetical-numerical order. In the seventeenth century, Antoine Rossignol vastly improved the basic design when he introduced a two-part system and scattered the elements so that they were not in alphabetical-numerical sequence. Instead, he prepared two lists: one with words in alphabetical order for encoding and the other in numerical sequence for decoding. In eighteenth- century Europe, codes of 1,500 to 3,000 numbered words were used to conceal especially sensitive data in diplomatic dispatches.

For example, Hugh Elliot, England's ambassador and espionage officer in Berlin during the American Revolution, used such a code when he transmitted confidential information to London after he stole Arthur Lee's secret journal. This confidential volume contained copies of American transactions with French and Spanish ministers; in addition, Lee lost copies of his personal correspondence when Elliot's agents entered Lee's quarters at the Berlin Inn in 1777. The ever-suspicious and exasperated Lee reported to Benjamin Franklin and Silas Deane, the American commissioners in Paris: "Public ministers have been regarded as spies; Mr. Elliot will give them the additional title of robbers."[1] Though the Prussian government officially accepted Elliot's false account that his eager servant stole the papers to please his master, King Frederick exclaimed in a private letter to his ambassador, Count Maltzan, in London, "Oh this worthy scholar of Bute, this incomparable man: your goddam Elliot. In truth, Englishmen ought to blush for shame that they sent such ambassadors to a foreign court."[2] Elliott was publicly rebuked by his government; however, his superior quietly awarded him £1000.[3]

For spies based in New York, the American military would also incorporate a special code during the revolution. Major Benjamin Tallmadge, General George Washington's director of secret service for five years after 1778, designed a one-part code of 763 elements. This young officer, born at Brookhaven, New York, in 1754, had been tutored by his father before enrolling at Yale University. Following graduation at age 19, Tallmadge became a high school superintendent in Connecticut before joining that state's regiment as a lieutenant a few weeks before the Declaration of Independence.

In 1778 Tallmadge began a secret spy ring in New York. As he modestly wrote later, "This year I opened a private correspondence with some persons in New York (for Gen. Washington) which lasted through the war. How beneficial it was to the Commander-in-Chief is evidenced by his continuing the same to the close of the war. I kept one or more boats continually employed in crossing the Sound on this business."[4] His secret agent was Robert Townsend, a store owner in New York, partner in a coffee house, and reporter for the society pages of James Rivington's newspapers. Townsend (codename: Culper, Jr.) transmitted information he gathered from his conversations with British officers. Townsend's coded messages were carried by Austin Roe, an employee, to Aaron Woodhull (codename: Samuel Culper, Sr.), a resident of Setauket, Long Island, who in turn sent the messages across the sound through Caleb

Brewster to Tallmadge or Enoch Hale, brother of Nathan Hale.

Tallmadge designed the code before July 1779 by using *Entick's Spelling Dictionary*, taking the most frequently used words, numbering them in alphabetical-numerical order and adding a mixed alphabet for words and numbers not listed on the sheet. He also assigned sixteen numbers for key individuals, along with thirty-six different cities or geographical entities. Tallmadge kept the original code in his possession, forwarded one copy to Long Island and sent the third copy to General Washington.[5]

Major Benjamin Tallmadge

George Washington's alphabet code sheet

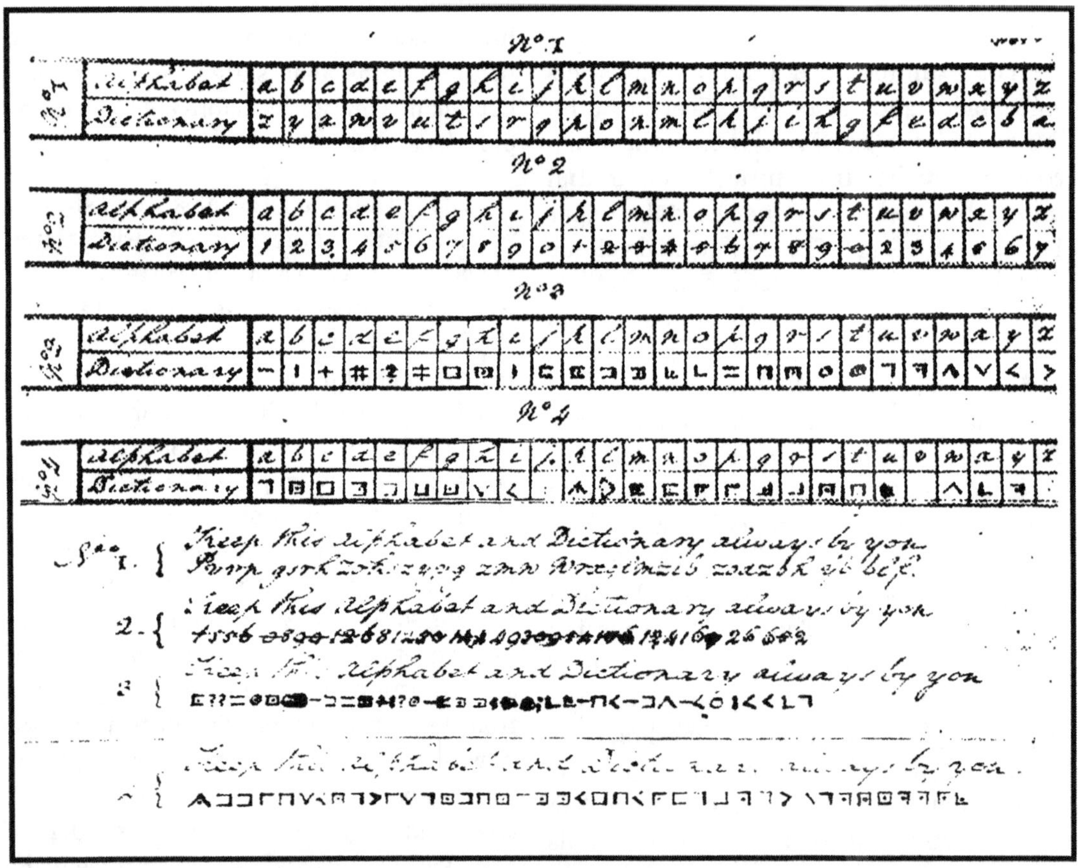

Noted below is an encoded letter that reveals the way the American spies used the code to disguise their messages:

This Tallmadge code, despite its inherent weaknesses due to alphabetical-numerical sequencing, provided an effective secret communications instrument, especially since it had codenumbers for twenty-two of the twenty-seven most frequently written words in the English language.

No. 21 729 [Setauket] 29 [August] 15th 1779

Sir Dqpeu [Jonas] beyocpu [Hawkins] agreeable to 28 [appoint] met 723 [Culper Junr.] not far from 727 [New York] & received a 356 [letter], but on his return was under the necessity to destroy the same, or be detected, but have the satisfaction to informe you that theres *[sic]* nothing of 317 [important] to 15 [advise] you of. Thers [*sic*] been no augmentation by 592 [ship] of 680 [war] or 347 [land] forces, and everything very quiet. Every 356 [letter] is opened at the entrance of 727 [New York] and every 371 [man] is searched, that for the future every 356 [letter] must be 691 [write] with the 286 [ink] received. They have some 345 [know] of the route our 356 [letter] takes. Ijudge it was mentioned in the 356 [letter] taken or they would not be so 660 [vigilant]. I do not think it will continue long so. I intend to visit 727 [New York] before long and think by the assistance of a 355 [lady] of my acquaintance, shall be able to out wit them all. The next 28 [appoint] for 725 [C. Brewster] to be here is the 1 of 616 [seventy, though Culper meant to write 617 for *August*] that it is so prolonged. It may be better times before then. I hope ther [*sic*] will be means found out for our deliverance. Nothing could induce me to be here but the ernest [*sic*] desire of 723 [Culper Junr.]. Friends are all well, and am your very Humble Servant, 722 [Culper Saml.][6]

Espionage Code

Source: Papers of George Washington, Library of Congress

1	a	37	attone	73	camp
2	an	38	attack	74	came
3	all	39	alarm	75	cost
4	at	40	action	76	corps
5	and	41	accomplish	77	change
6	art	42	apprehend	78	carry
7	arms	43	abatis	79	clergy
8	about	44	accommodate	80	common
9	above	45	alternative	81	consult
10	absent	46	artillery	82	contest
11	absurd	47	ammunition	83	contract
12	adorn	48	be	84	content
13	adopt	49	bay	85	Congress
14	adore	50	by	86	captain
15	advise	51	best	87	careful
16	adjust	52	but	88	city
17	adjourn	53	buy	89	clamour
18	afford	54	bring	90	column
19	aifrent	55	boat	91	copy
20	affair	56	barn	92	cover
21	again	57	banish	93	county
22	april	58	baker	94	courage
23	agent	59	battle	95	credit
24	alter	60	better	96	custom
25	ally	61	beacon	97	compute
26	any	62	behalf	98	conduct
27	appear	63	bitter	99	comply
28	appoint	64	bottom	100	confine
29	august	65	bounty	101	caution
30	approve	66	bondage	102	conquer
31	arrest	67	barron	103	coward
32	arraign	68	brigade	104	confess
33	amuse	69	business	105	convict
34	assign	70	battery	106	cannon
35	assume	71	battalion	107	character
36	attempt	72	british	108	circumstance
109	clothier	156	deliver	203	forget
110	company	157	desolate	204	fulfil[*sic*]

111	confident	158	during	205	factor
112	committee	159	ear	206	faculty
113	continue	160	eye	207	favorite
114	contradict	161	end	208	fortune
115	correspond	162	enquire	209	forget
116	controversy	163	effect	210	foreigner
117	commission	164	endure	211	fortitude
118	commissioner	165	enforce	212	fortify
119	constitution	166	engage	213	formiable
120	date	167	enclose	214	foundation
121	day	168	equip	215	february
122	dead	169	excuse	216	get
123	do	170	exert	217	great
124	die	171	expend	218	good
125	damage	172	expose	219	gun
126	doctor	173	extort	220	go
127	duty	174	express	221	gain
128	drummer	175	embark	222	guide
129	daily	176	employ	223	gold
130	dispatch	177	explore	224	glory
131	distant	178	enemy	225	gunner
132	danger	179	example	226	gloomy
133	dislodge	180	embassador	227	govern
134	dismiss	181	engagement	228	grandieure
135	dragoons	182	experience	229	guilty
136	detain	183	evacuate	230	guinea
137	divert	184	Farm	231	gallant
138	discourse	185	face	232	gazette
139	disband	186	fate	233	grateful
140	dismount	187	false	234	glacis
141	disarm	188	friend	235	general
142	detect	189	fin	236	garrison
143	defense	190	find	237	gentleman
144	deceive	191	form	238	glorious
145	delay	192	fort	239	gradual
146	difficult	193	fleet	240	granadier
147	disapprove	194	famine	241	hay
148	disregard	195	father	242	he
149	disappoint	196	foggy	243	his
150	disagree	197	folly	244	him
151	disorder	198	frugal	245	haste
152	dishonest	199	faithful	246	hand

153	discover	200	favour	247	hang
154	december	201	faulty	248	has
155	demolish	202	foreign	249	have
250	head	297	insnare	344	kill
251	high	298	instruct	345	know
252	hill	299	intrigue	346	law
253	hope	300	intrust	347	land
254	hut	301	instant	348	love
255	horse	302	invest	349	low
256	house	303	invite	350	lot
257	happy	304	ignorant	351	lord
258	hardy	305	impudent	352	light
259	harvest	306	industry	353	lart
260	horrid	307	infamous	354	learn
261	horseman	308	influence	355	lady
262	human	309	infantry	356	letter
263	havock	310	infantry	357	levy
264	healthy	311	injury	358	levies-new
265	heavy	312	innocent	359	liar
266	honest	313	instrument	360	lucky
267	hunger	314	intimate	361	language
268	honor	315	illegal	362	limit
269	harmony	316	imagin	363	liquid
270	hazardous	317	important	364	longitude
271	hesitate	318	imprison	365	latitude
272	history	319	improper	366	laudable
273	horrible	320	incumber	367	legible
274	hospital	321	inhuman	368	liberty
275	hurrican [*sic*]	322	inquiry	369	lottery
276	hypocrite	323	interview	370	literature
277	[?]	324	incorrect	371	man
278	[?]	325	interceed	372	map
279	[?]	326	interfere	373	may
280	I	327	intermix	374	march
281	if	328	introduce	375	mast
282	in	329	immediate	376	make
283	is	330	impatient	377	met
284	it	331	incouragemt	378	me
285	ice	332	infection	379	my
286	ink	333	irregular	380	much
287	into	334	invalid	381	move
288	instance	335	indians	382	mort

289	island	336	june	383	mine
290	impress	337	july	384	many
291	improve	338	jury	385	mercy
292	incamp	339	jealous	386	moment
293	incur	340	justify	387	murder
294	infest	341	january	388	measure
295	inforce	342	key	389	method
296	instance	343	king	390	mischief
391	mistake	438	onset	485	purpose
392	molest	439	order	486	people
393	majesty	440	over	487	pleasure
394	meditate	441	obstruct	488	produce
395	memory	442	obtain	489	prison
396	messanger	443	observe	490	progress
397	misery	444	occur	491	promise
398	moveable	445	offense	492	proper
399	multitude	446	omit	493	prosper
400	miscarry	447	oppose	494	prospect
401	misfortune	448	obligate	495	punish
402	miserable	449	obstinate	496	pertake
403	mercenary	450	obviate	497	perform
404	majority	451	occupy	498	permit
405	minority	452	operate	499	pervert
406	memorial	453	origin	500	prepare
407	missterious [*sic*]	545	ornament	501	prevail
408	manufacture	455	overcome	502	preserve
409	moderator	456	overlook	503	pretend
410	minsterial	457	overtake	504	promote
411	name	458	overrun	505	propose
412	new	459	overthrow	506	protect
413	no	460	obediance	507	provost
414	not	461	objection	508	pursue
415	night	462	october	509	passenger
416	never	463	obscure	510	passion
417	needful	464	occasion	511	pension
418	number	465	opinion	512	period
419	neither	466	oppression	513	persecute
420	nothing	467	opportunity	514	poverty
421	neglect	468	obligation	515	power or
422	nation	469	pay		powerful
423	navy	470	peace	516	prosperous
424	natural	471	plan	517	punishment
425	negative	472	put	518	preferment

426	negligence	473	port	519	production
427	novembr	474	proof	520	pursuant
428	necessary	475	please	521	pensioner
429	nobility	476	part	522	Parliament
430	oath	477	paper	523	persecution
431	of	478	pardon	524	practicable
432	off	479	party	525	profitable
433	on	480	perfect	526	particular
434	or	481	pilot	527	petition
435	out	482	prudent	528	profession
436	offer	483	publish	529	proclaim
437	office	484	purchase	530	provision
531	Protection	578	remember	625	the
532	quick	579	remittance	626	that
533	question	580	represent	627	this
534	quantity	581	rebellion	628	these
535	quallity [*sic*]	582	reduction	629	they
536	rash	583	remarkable	630	there
537	rain	584	reinforcement	631	thing
538	run	585	refugee	632	though
539	rule	586	sail	633	time
540	read	587	see	634	to
541	rise	588	sea	635	troops
542	random	589	scheme	636	thankfull [*sic*]
543	ransom	590	set	637	therefore
544	rather	591	send	638	timber
545	real	592	ship	639	tory
546	riot	593	safe	640	transport
547	robber	594	same	641	trail
548	ready	595	sky	642	traitor
549	ruin	596	secret	643	transgress
550	ruler	597	seldom	644	translate
551	rapid	598	sentence	645	terrible
552	reader	599	servant	646	tyranny
553	rebel	600	signal	647	vain
554	rigor	601	silent	648	vaunt
555	river	602	suffer	649	vouch
556	receit	603	sudden	650	vacant
557	refit	604	surprise	651	vary
558	regain	605	summer	652	venture
559	rejoice	606	speaker	653	vital
560	relate	607	steady	654	vulgar

561	request	608	submit	655	value
562	relax	609	surpass	656	virtue
563	redoubt	610	sanction	657	visit
564	rely	611	sensible	658	valiant
565	remit	612	singular	659	victory
566	reprieve	613	soldier	660	vigilant
567	repulse	614	sovereign	661	vigorous
568	reward	615	security	662	violent
569	retract	616	seventy	663	volunteer
570	resign	617	August	664	valuable
571	ratify	618	september	665	voluntary
572	recompence	619	surrender	666	up
573	regular	620	serviceable	667	upper
574	regulate	621	security	668	upon
575	rigorous	622	severity	669	unto
576	recital	623	society	670	unarm
577	recover	624	superior	671	unfit
672	unheard	719	North, Lord		
673	unsafe	720	Germain, Lord		
674	uniform	721	Bolton John		
675	uncertain	722	Culper Sam.l		
676	uncommon	723	Culper Junr.		
677	unfriendly	724	Austin Roe		
678	unfortunate	725	C. Brewster		
679	wind	726	Rivington		
680	war	727	New York		
681	was	728	Long Island		
682	we	729	Setauket		
683	will	730	Kingsbridge		
684	with	731	Bergen		
685	when	732	Staten Island		
686	wharf	733	Boston		
687	wound	734	Rhode Island		
688	wood	735	Connecticut		
689	want	736	New Jersey		
690	wait	737	Pensylvania [*sic*]		
691	write	738	Maryland		
692	who	739	Virginia		
693	wish	740	North Carolina		
694	whose	741	South Carolina		
695	wages	742	Georgia		
696	warlike	743	Quebeck		

697	welfare	744	Hallifax
698	willing	745	England
699	winter	746	London
700	water	747	Portsmouth
701	woman	748	Plymouth
702	writer	749	Ireland
703	waggon [*sic*]	750	Corke
704	weary	751	Scotland
705	warrant	752	West Indies
706	yet	753	East Indies
707	you	754	Gibralter
708	your	755	France
709	yesterday	756	Spain
710	zeal	757	Scotland
711	Gen[l] Washingtin	758	Portugal
712	Gen[l] Clinton	759	Denmark
713	Tryon	760	Russia
714	Erskine	761	Germany
715	Vaughan	762	Hanover
716	Robinson	763	Head Quarters
717	Brown		
718	Gen[l] Garth		

Alphabet

a	e
b	f
c	g
d	h
e	i
f	j
g	a
h	b
i	c
j	d
k	o
n	p
o	q
p	r
q	k
r	l
s	u
t	v
u	w
v	x
w	y
x	z
y	s
z	t

Numbers

1	e
2	f
3	g
4	i
5	k
6	m
7	n
8	o
9	q
0	u
1	m

Directions for the Alphabet

N.B. The use of this Alphabet is when you wish to express some words not mentioned in the numerical Dictionary. For instance the word *heart.* would be expressed this *bielv.* look the [sic] letters of the real world in the first column of the alphabet and then opposite to them, let those letters in the second column to draw a line under the word, as *fwv* stands for but. Numbers are represented by theri opposite letters which must have a double line under them as *fikm* is 2456. & *nqu is 790....*

Directions for the numerical Dictionary. In the numerical Dictionary it is sufficient to express a part of the sentence only in figures, to make the rest perfectly unintelligible, as all words cannot be mentioned those of synonimous meaning must be sought for. & if ot to be found, & the word not proper to be wrote, then the alphabet must be used – When numbers are used always observe to put a period after the number thus 284. stands for *it & 295. inforce.* It will often hapen that the same word may need t be changed thro the different moods, tenses, numbers. Thus if you would be 328. if you would express the word introduced make a small flourish over the same 328. Horse is repres. by 255. Horses by 255, kill by 344. killed by 344. impress by 290. impressed by 290, in such cases the fore going & subsequent parts must determine the word.

Notes

1. Arthur Lee to Franklin and Deane, Berlin, 28 June 1777, in Richard Henry Lee, *Life of Arthur Lee, LL.D.* (Boston: Wells and Lilly, 1829), 2:10. In fact, Elliot soon realized that the only significant information he had stolen was the draft of a treaty with Spain and a letter from Frederick II which suggested eventual recognition of the United States. The other information had previously been reported by the spy in Franklin's office, Edward Bancroft, and other British spies. Cf. Louis Watson Potts, "Arthur Lee - American Revolutionary." Ph.D. dissertation, Duke University, 1970,271-75.

2. As quoted in Oscar Browning, "Hugh Elliot in Berlin," in Transactions of the Royal Historical Society, New Series, 4 (1889), 87-88. Thomas Carlyle's account in the *History of Frederick II of Prussia called Frederick the Great* (Boston: Estes and Lauriat, 1885), 7:384-386 contains a very inaccurate account of the robbery: though the almost fictional account is fascinating, his facts are wrong. For example, he wrote that a score or more writers copied the documents. George Bancroft's one paragraph summary merely recorded that the papers were stolen, read by Elliot, and then returned. The historian included part of Frederick's letter to his minister in London but omitted Frederick's reference to damnation. Cf. *History of the United States from the Discovery of the Continent* (New York: D. Appleton & Co., 1895, 5:240. Richard Henry Lee's account in Lee, Life, 1: 95-96, includes several inaccuracies: entry into Lee's room was through a window rather than a door; the papers were returned before a formal inquiry was made. Years later, Elliot tried to convince John Quincy Adams of his innocence.

3. Louis W. Potts, Arthur Lee: *A Virtuous Revolutionary* (Baton Rouge: Louisiana State University Press, 1981), 176.

4. *Memoir of Colonel Benjamin Tallmadge* (New York, 1858; reprinted by Arno Press, 1968), 29.

`5. Tallmadge to Washington, n.p., 24 July 1779, as cited in Charles Swain Hall, *Benjamin Tallmadge: Revolutionary Soldier and American Businessman* (New York: AMS Press, Inc., 1966), 50-51; also, Morton Pennypacker, *General Washington's Spies on Long Island and In New York* (Brooklyn: Long Island Historical Society, 1939), 209. There has been considerable doubt about the Loyalist or Patriot support given by James Rivington, the publisher, who may have served as double agent. The best review of the evidence may be found in Catherine Snell Crary, "The Tory and the Spy: The Double Life of James Rivington," *The William and Mary Quarterly*, 16 (January, 1959),61-72.

6. Pennypacker, *Washington's Spies*, 252-253. The words in brackets have been supplied from the Tallmadge code.

Chapter 5

Dictionary Codes

"We must fall on some scheme of communicating our thoughts to each other, which shall be totally unintelligible to every one but to ourselves."[1] In 1764, Thomas Jefferson, then twenty years old, and some twelve years before he would craft the Declaration of Independence, wrote from Williamsburg to his dear friend, the Virginia planter and legislator John Page. He lamented the lack of security for carrying on confidential correspondence, and especially the difficulties in hiding information regarding a young lady he was courting. He explained to Page that he would use Thomas Shelton's *TachyGraphy, The Most Exact and Compendious Met hode of Short and Swift Writing That Hath Ever Been Published*, published over a century earlier, in London. Specific instructions, he wrote, would follow. Jefferson's anxieties about keeping his correspondence entirely private centered on his strong desires to hide his eager courtship of Rebecca Burwell. In earlier letters to Page, he disguised her name by referring to her as "R.B.," "Belinda," "Adnileb," and "Campana in die." But these terms did not offer sufficient protection against those who might open his letters and pry into his personal affairs.

Thus in his earliest correspondence, the future United States president, indeed, a president who was the most prolific of all his predecessors and successors in using codes and ciphers and also designing new ones, recognized the need for secure communication and sought a readily available method for encoding his letters. This American Founding Father sought privacy, as did numerous others.

Of all the foreign ministers from America during the American Revolution, only John Jay, even before he arrived at his post in Spain, exhibited the most interest and offered the greatest originality in basing codes upon books for masking correspondence. Born in New York City in 1745, educated by private tutors and then at King's College, Jay continued his law studies in the city and was admitted to the bar in 1768. As a delegate to the First and Second Continental Congresses, he became convinced of the necessity for American independence. Chief Justice of New York, member of the Continental Congress, member of the Secret Committee of the Continental Congress for Corresponding with Foreign Nations, Jay became a powerful statesman in the struggle for independence. In 1779, and less than ten months after being elected president of the Congress, Jay was chosen as minister plenipotentiary to Spain. The Congress hoped his prestige and talented negotiating skills would win Spanish military support and additional funds for the Revolution.

After a miserable voyage across the stormy Atlantic Ocean with his family, and soon after his arrival in Spain, where mail opening by government spies was common, Jay wrote to Robert Livingston and suggested that the second part of Abel Boyer's *French Dictionary*, 13th edition, printed in London in 1771 (in which the English section was placed before the French), be used. Since the book was not paged, Jay asked that this be promptly done, with the first page numbered 1 and so on. In addition, as there were three columns on each page, the first column should be noted as "c" and second, "a," and third, "b." The last element in the instructions specified Livingston should count the number of words from the top, including the word he planned to use, and add 7 to the number. Hence, according to the example that Jay provided, the word abject was the third word in the third column of page 2, and therefore the code number would be "2 b 10." For his correspondence with Charles Thomson, secretary of the Continental Congress, Jay added 5 to the number of the page that was to be listed last.[2] Jay's clever secret design failed in one instance, however, because Livingston could not acquire a copy of Boyer's book.

In Madrid, Jay also chose another book for encoding his correspondence, *Entick's Spelling Dictionary*: *The new spelling dictionary to which is prefixed, a grammatical introduction to the English tongue*, first published in London in 1765, printed in 1777. In his correspondence with the wealthy Philadelphia merchant and banker Robert Morris, Jay specified that Morris should page the book backwards with the last page of the book, 468, to be made page 1, and the title page, 468. Moreover, words on the page should be counted from the top and the columns distinguished by a dot over the first figure for the first column and a dot over the second figure for the second column. For example, the word absent was the fifth word in the first column of page 434, and therefore the code would be written 5.434.[3]

For another correspondent, William Bingham, a founder of the Bank of North America and prosperous Philadelphia trader, Jay also sent instructions keyed to Entick's; however, he instructed Bingham to add 20 to the number of the page and 10 to that of the word used. Dots to indicate columns were also added along with a simple substitution alphabet for names or words not in the dictionary.[4]

Certain British leaders also used Entick's for their own code systems in America. James Lovell, creator of numerous American cipher systems for diplomatic correspondence, suspected this and analyzed several intercepted British military letters in October 1781 and found they were encoded in accord with this dictionary. The British, however, simply listed the page, column, and the word, such as 115.1.4.

Arthur Lee, Virginia born, educated at Eton, later at the University of Edinburgh in medicine, and finally at Lincoln's Inn and the Middle Temple in law, became fascinated by America's prospects for a successful revolution. In 1770, he became agent for Massachusetts in London and actively protested British regulatory measures against the American colonies. In late 1775, the Secret Committee on Foreign Correspondence of the Continental Congress, asked him to become its confidential correspondent, and one year later he was appointed, along with Benjamin Franklin and Silas Deane, to become a commissioner to the court of France.

Writing to the committee shortly before the signing of the Declaration of Independence, Lee suggested the following plan for encoding dispatches: "This book is better than the last I sent you. It is to decypher what I wrote to you & for you to write by. This is done by putting the page where the word is to be found and the letter of the alphabet corresponding in order with the word. As there are more words in a page than the letter of the alphabet the letter must be doubled or trebled to answer that. As thus, to express the troops: you write 369, kk 381, vv- ing, ed, s, & must be added when necessary, and distinguished by making no comma between them and the figures. Thus, for betray'd, put 33 ed x. The letters I use are abcdefghijklmnopqrstuvw w.ch are 26. I cant use this till I know it is safe. You can write to Mrs. Lee on Tower Hill in a woman's hand. If you have both books say the children are well: if the first only, the eldest child is well, if this, the youngest child is well. They will let this pass."[5]

The most complete set of instructions for a book code in America's early years was prepared by Thomas Jefferson and James Madison in January 1783. Based upon *Thomas Nugent's New Pocket Dictionary of the French and English Languages* (London, Dilly, 1774), this special code, according to Jefferson, included rather complicated guidelines:

> The 1st number denotes the column — the 2d the line, to avoid a 3d number never use but the 1st word of the line.
>
> A line which goes across more than one column is not to be counted, because it cannot be said to belong to one column more than another. e.g. The 5 upper lines over columns 949.950 are not to be counted.
>
> One or more letters belonging to a column are counted as a line. e.g. A B is 1.6 or 2.6 or 3.6. Abaddon is 1.40. B Y is 132.6. C is 133.18. Ecluse [*sic*] or Shuy [*sic*] is 1064.26 one or more syllables or an &c. if they occasion or occupy a line is to be counted as such. e.g. Mich is 1068.17.
>
> Nouns are pluralized or genitived by this mark over them 'letters are doubled by a comma under them. a, following a verb denotes its participle active. p, its participle passive. thus 137.39a is buying. l32.39p is bought. the part. pass. may be used for the indic. imperf. the person will be known by the pronoun prefixed.. e.g. 402.5 132.39. is he buy. 402.5 l32.39p. is he bought.

Furthermore, Jefferson wrote, "Numbers are to be written as the words that express them. e.g. 42.4 43.3 is twenty-two. Frequently throw in higher numbers than 1545 which meaning nothing will serve to perplex." However, Jefferson would add 8 numbers above 1545 to designate certain individuals in France, such as Dr. Franklin, who was "1885."[6]

Madison spent hours trying, frequently unsuccessfully, to decode Jefferson's correspondence, for the author sometimes miscounted the lines, or the addressee misunderstood Jefferson's instructions for this complicated code system. Jefferson also had problems, and the encoded Jefferson-Madison correspondence presents a covey of garbled messages for future editors to translate into plain text. In frustration, the two colleagues set aside the dictionary code book, containing such detailed additives, and turned to code lists that, while not error-proof, did eliminate the common error of miscounting lines and word positions. And though dictionaries would occasionally be used in the succeeding decades by Americans, code sheets became far more common, and indeed, more error-proof, especially for military and diplomatic correspondence.

Notes

1. Thomas Jefferson to John Page, Devilsburgh, 19 January 1764, in Julian P. Boyd, *The Papers of Thomas Jefferson* (Princeton: Princeton University Press, 1950), 1:15.
2. John Jay to Robert Livingston, Cadiz, 19 February 1780 in Richard B. Morris, "The Jay Papers: Mission to Spain," *American Heritage*, 19 (February 1968), 85; also, Richard B. Morris, ed., *John Jay: The Making of a Revolutionary, Unpublished Papers 1745-1780* (New York: Harper & Row, 1975), 735-737. Jay used *Boyer's French Dictionary* for encoding a crucial paragraph in a letter to Samuel Huntington, Madrid, 6 November 1780: cf. ibid., 1:829-830. Also cf. Jay to Thomson, Cadiz, 29 February 1780 in *Jay Papers*, Columbia University.
3. John Jay to Robert Morris, Madrid, 19 November 1780 in Henry P. Johnston, ed., *The Correspondence and Public Papers of John Jay* (New York: Putnam, 1890), 1:445. Also cf. Morris to Jay, Philadelphia, 5 July 1781, in E. James Ferguson, ed., *The Papers of Robert Morris* (Pittsburgh: University of Pittsburgh Press, 1973-1975), 1:113.

4. John Jay to William Bingham, St. Ildefonso, 8 September 1781 in Johnston, Correspondence, 2:66-69.

5. Arthur Lee to the Secret Committee, 3 June 1776 in *Papers of the Continental Congress*, Record Group 360, Microcopy 247, Roll 110.

6. This description is found in the *Continental Congress Papers*, Virginia State Library. Jefferson would first use the code in a letter to James Madison, Baltimore, 31 January 1788, in Boyd, *Jefferson Papers*, 6:225-226. Considerable confusion surrounds this particular edition of *Nugent's Dictionary* since apparently no copy has been located in the United States, and, therefore, interpolation has been practiced by previous editors. The editors of *The Papers of James Madison*, William T. Hutchinson and William M.E. Rachal, were assisted by the Honorable J. Rives Childs, a U.S. Army cryptanalyst in World War I who found the 1774 edition in the Bibliotheque Nationale in Paris. These editors noted that this edition had no pagination, and, therefore, Jefferson and Madison must have numbered their pages in a special way. For example, pages 1 through 64 were numbered in consecutive sequence; however, beginning with the next page, 15 is added and the page number becomes 80; consecutive page 72 is raised by 23 to become 95; page 103 is raised by 29 to become 132, and so on, until page 903 became 1028. Cf. Hutchinson and Rachal, eds., *Papers of James Madison* (Chicago: University of Chicago Press, 1969), 6:177ff. Other early editions found at the Bibliotheque Nationale in Paris are 1767, 1779, 1784,1786, and 1826.

Chapter 6

General George Washington's Tradecraft

> **The necessity of procuring good Intelligence is apparent and need not be further urged. All that remains for me to add is, that you keep the whole matter as secret as possible. For upon secrecy, success depends in Most Enterprises of the kind, and for want of it, they are generally defeated, however well planned and promising a favorable issue.[1]**

These closing comments from General Washington in a 1777 dispatch to Colonel Elias Dayton brilliantly summarized the instructions Washington had given Dayton: obtain secretly the enemy strength on Staten Island, together with the location and strength of their guards. And little more than a year later, Washington, like a case officer for a modern intelligence agency, instructed his director of the secret service, Major Benjamin Tallmadge, on the chief elements for gathering and reporting intelligence information; and hoped these ideas would be implemented by his special spies, codenamed "Culper," who were under Tallmadge's control.

> **As all great movements, and the fountain of all intelligence must originate at, and proceed from the head Quarters of the enemy's army, C________ had better reside at New York, mix with, and put on the airs of a Tory to cover his real character, and avoid suspicion. In all his communications he should be careful in distinguishing matters of fact, from matters of report. Reports and actions should be compared before conclusions are drawn, to prevent as much as possible, deception.**

Continuing on, Washington noted special attention should be given to the arrival and departure of naval vessels, movement and destination of troops, sizes of reinforcement, numbers of recruits for filling out the regiments. And the general wanted careful reports on the milieu of the times: "The temper and expectation of the Tories and Refugees is worthy of consideration, as much may be gathered from their expectations and prospects; for this purpose an intimacy with some well informed Refugee may be political and advantageous." Moreover, Washington continued, it would be wise "to contract an acquaintance with a person in the Naval department, who may either be engaged in the business of providing Transports for the embarkation of the Troops, or in victuelling [*sic*] of them."[2]

In another set of instructions, Washington urged that his informant in New York City mix among the British officers and refugees in the coffee houses and other public places: to learn how their transports were protected against attack, whether by chains or booms to ward off fire rafts, or by armed ships. Also, he asked about the harbor fortifications, number and size of cannon, whether there were pits dug within and before defensive lines, and whether they were three or four feet deep and had sharp pointed stakes installed that were intended to wound men who attempted a night attack.[3]

Besides advice, Washington also provided the Culper spies with invisible ink – he called it the "white ink" – for their dispatches. The use of secret fluids dates back to antiquity. Writing in the generation after Julius Caesar's reign, the poet Ovid revealed in *The Art of Love:* "A letter too is safe and escapes the eye when written in new milk:

touch it with coal dust and you will read. That too will deceive which is written with a stalk of moistened flax, and a pure sheet will bear hidden marks."[4]

General Washington explained to Tallmadge that he was sending all the special writing chemical that he had in phial number 1. In phial 2 was the liquid that made the white ink visible by wetting the paper with a brush. Utmost secrecy regarding these materials, Washington told Tallmadge, was indispensable. Sir James Jay, a London physician and John Jay's brother, invented the two special fluids and sent a supply to his brother and also General Washington. Early in the revolution, Sir James used the ink at the bottom of brief friendly letters to his brother and told him of the British ministry's decision to force the colonies into submission; he also wrote from London to Franklin and Deane in Paris and warned them of General John Burgoyne's intended invasion from Canada. Silas Deane had been given a supply of the precious ink by John Jay shortly before sailing for France in March 1776, and later James Jay sent additional supplies. Robert Morris told John Jay to apply the special ink on his letters to Deane.

In September 1779, Washington wrote from his headquarters at West Point and taught his director of secret service several practical techniques for Culper to use for hiding the secret messages, thus protecting the messengers who carried the dispatches. Culper should "occasionally write his information on the blank leaves of a pamphlet; on the first second &c. pages of a common pocket book; on the blank leaves at such end of registers almanacks [*sic*] or any new publication or book of small value. He should be determined in the choice of these books principally by the goodness of the blank paper, as the ink is not easily legible, unless it is on paper of good quality. Having settled a plan of this kind with his friend, he may forward them without risque of search or the scrutiny of the enemy as this is chiefly directed against paper made up in the form of letters."[5]

Another method would be, Washington wrote, to write a letter on domestic affairs to his friend at Setauket, Long Island, and write with invisible ink between the lines, or on the opposite side of the page. To distinguish these letters meant for Washington, Culper could leave off the place or date (putting the date in invisible ink), or perhaps fold letter in a special manner. However, he concluded that the mode of writing in the fly leaves of books seemed the safest method to him. Perhaps recalling the capture and imprisonment of Benjamin Church, who aroused suspicions with his encoded letter in 1775, Culper apparently preferred using the invisible ink rather than the code sheet prepared by Tallmadge. Moreover, the British troops were opening all letters carried to New York, and an encoded letter always raised suspicions. His New York spies must have used the secret inks because Washington continued to urge his agents to economize in writing with the special ink because he had only small amounts.

With growing frustration, Washington wrote to Tallmadge in February 1780 and sent hard money of twenty guineas and more invisible ink for the New York agents. Culper was using the ink on blank sheets of paper and sending them by messenger: this, complained the general, was bound to raise suspicions. Rather, Tallmadge should tell him again to write in "Tory stile," describing family matters, and between the lines write in invisible ink the special intelligence information.[6]

General Washington knew how to obtain special intelligence and, as importantly, how to mask it in dispatches. His thoughtful, thorough, and creative instructions reflected experience and practical knowledge on espionage practices, and especially secret writing. Facing an enemy that had overwhelming military power, Washington recognized the crucial necessity for intelligence and secrecy, for they promised military success and, as well, the continued independence of a new nation.

Notes

1. General George Washington to Colonel Elias Dayton, 8 Miles East of Morris Town, 26 July 1777, in John C. Fitzpatrick, ed., *The Writings of George Washington from the Original Manuscript Sources, 1745-1799* (Washington, D.C.: Government Printing Office, 1933), 8:479.

2. General George Washington to Major Benjamin Tallmadge, Middlebrook, 21 March 1779, in ibid., 14:277.

3. Instructions for C Senior and C Junior, [14 October 1779], in ibid., 16:466.

4. As quoted in Rose Mary Sheldon, *Tinker, Tailor, Caesar, Spy: Espionage in Ancient Rome* (Ann Arbor: UMI Dissertation Information Service, 1987), 268. This delightful and thorough study of ancient espionage practices reports "evidence for codes and ciphers is meager," 266. She also notes a secret writing system used by Hannibal's father, Hamilcar Barca: he "wrote messages on a wooden tablet and then covered it with fresh wax to look like a blank message board (normally the message was inscribed in the wax). But the practice long antedates Hamilcar's use of it; it was common for the Romans to ascribe the invention of clever stratagems to their defeated enemies," 28.

5. General George Washington to Major Benjamin Tallmadge, West Point, 24 September 1779, in Fitzpatrick, Writings, 16:331.

6. General George Washington to Major Benjamin Tallmadge, Morristown, 5 February 1780, in ibid., 17:493.

Chapter 7

American Postal Intercepts

"Whereas the communication of intelligence with regularity and despatch, from one part to another of these United States, is essentially requisite to the safety as well as the commercial interest thereof . . . ." The wartime Congress under the Articles of Confederation established an ordinance for regulating the Post Office in October 1782.[1] Under the jurisdiction of the postmaster general, a series of posts would be established and maintained from the state of New Hampshire down through the state of Georgia, and other areas chosen by him or the Congress.

Further, the Congress stated in very precise legal terms, "the Postmaster General, his clerk or assistant, his deputies, and post and express-riders, and messengers, or either of them, shall not knowingly or willingly open, detain, delay, secrete, embezzle or destroy, or cause, procure, permit or suffer to be opened, detained, delayed, secreted, embezzled or destroyed any letter or letters, packet or packets, or other despatch or despatches, which shall come into his power, hands or custody by reason of his employment in or relating to the Post Office, except by the consent of the person or persons by or to whom the same shall be delivered or directed. . . "

The only other procedures permitting interference with the mail were by "an express warrant under the hand of the President of the Congress of these United States, or in time of war, of the Commander in Chief of the armies of these United States, or of the commanding officer of a separate army in these United States, or of the chief executive officer of one of the said states, for that purpose, or except in such other cases wherein he shall be authorised to do by this ordinance. . ."

However, especially anxious to protect the confederation government's mail from state interference, the congressional authorities ordinance provided "no letter, franked by any person authorised by this ordinance to frank the same, shall be opened by order of any military officer, or chief executive officer of either of the states."[2] Those with the franking privilege included members and secretary of the Congress while in attendance in Congress, to and from the commander in chief of the United States armies, commander of a separate army, to and from the heads of the departments of war, finance, and of foreign affairs on public service and finally any officers of the line on active duty.

Penalties for breaking these rules were severe: the postmaster general could be fined $1,000 (his annual salary was $1,500), and postmasters, post-riders, and others employed in the Post Office Department, $300. Moreover, persons found guilty of this crime could never hold any office of trust or profit in the United States.

Although Great Britain and the United States signed a treaty of peace at Paris in September 1783, and General Washington ordered his army disbanded in November, the transition of the nation to peacetime moved slowly and cautiously. Though winning independence, the Confederation Congress remained suspicious about the British, particularly their willingness to surrender Northwest forts or posts as promised in the treaty of peace. And as well, American governmental leaders became troubled by British attempts, real or imagined, to detach frontier areas from the weakened United States.

In early September 1785, John Jay, secretary of foreign affairs, wrote to the president of Congress about his deep concerns regarding British intentions. This dispatch also reveals the uneasiness and suspicions in postwar America. An obviously distressed Jay wrote,

> **The English Packet which arrived the Day before Yesterday brought me no Letters from Mr. Adams; which Impute to its being a Mode of Conveyance to which nothing very important can prudently be trusted.**
>
> **Some private Intelligence by that Vessel leads me to consider the Surrender of our Posts as being more problematical than it has lately appeared to be.**
>
> **I hear that the Circumstance of Congress having ordered some Troops to be raised, excited the Attention of the british [*sic*] Ministry, and induced them to order two Regiments to embark for Quebec — a Packet was preparing to sail for that Place on the first Wednesday in last Month with Despatches, which was perhaps it was not thought expedient to convey there through our Country.**
>
> **The Loyalists at the different Posts are computed to amount to between six and seven thousand and I am assured that they are provided with Arms and Rations by Government.**
>
> **. . .What Degree of Credit is due to this Intelligence is not in my Power to ascertain. It nevertheless comports with certain Reports which have lately reached us from the Frontiers, Vizt. that Encouragement was given by the Government of Canada to our People to settle Lands in the vicinity of the Posts; and that a considerable Number of Persons from among us had been seduced by their Offers to remove thither.**
>
> **I think it my Duty to lay these Matters before Congress and at the same Time to observe that in my Opinion they should for the present be kept secret.**

And then Jay added a suggestion for peacetime postal letter opening, the first in the nation's history: "Permit me Sir, also to hint, that there may be Occasions when it would be for the Public Interest to subject the Post Office to the Orders of your principal executive Officers."[3]

Secretary Jay probably recalled the traditional English practice of mail opening that dated back to the Middle Ages. The British Post Office came into being to carry the monarch's letters, and gradually when citizens also used it, the king claimed the right to examine their letters. Oliver Cromwell organized the process more carefully by having his own officer in a special room at the Post Office open, read, and reseal suspicious documents each evening between 11 and 4. By 1711, warrants were required, signed by a principal secretary of state before the letter could be opened. A brief time later, the secret room was divided into two parts: a "Secret Office" for handling foreign correspondence and a "Private Office" for domestic letters. A deciphering branch became an integral division.[4]

Jay's request for letter-opening powers was ordered to be kept secret, and the Congress promptly acted on his suggestion two days after he sent the message. With representatives from all thirteen states present, delegate Charles Pinckney of South Carolina, who had been captured during the revolution by the British at Charlestown, made the following motion: "Resolved, That, whenever it shall appear to the Secretary of the United States of America for the department of foreign affairs that their safety or interest require the inspection of any letters in any of the post Offices, he be authorized and empowered to inspect the said letters, excepting from the operation of this resolution, [which was to continue for the term of twelve months] all letters franked by or addressed to members of Congress."[5]

Passed by the Congress, this resolution was entered only in the Secret Journal, Domestic, of the Congress. One year later, the Congress again considered the authority and resolved unanimously "That whenever it shall appear to the Secretary of the United States of America for the department of foreign Affairs that their safety or interest require the inspection of any letters in any of the post Offices he be authorised and empowered to inspect the said letters, excepting from the operation of this resolution all letters franked by or addressed to Members of Congress."[6] Moreover, it was noted by Charles Thomson, the careful and thorough secretary of the Congress, that the renewal was passed without a limitation of time.[7]

When Congress under the Constitution passed legislation for the temporary establishment of the post office in 1789, it specified that the regulations would be the same as they had been under resolutions and ordinances of the previous Congress. In 1792, when the Congress passed permanent and extensive regulations for the postal system, it specified that if any post office employee should open the mail, or destroy letters unrelated to money, and be found guilty, the penalty would not exceed $300 or imprisonment for more than six months. However, if an employee took a letter that contained bank notes or other forms of money such as bonds or promissory notes for the payment of money, he would suffer the penalty of death if found guilty. A similar death penalty awaited anyone found guilty of robbing the mail carrier.[8] Further legislation in 1794 continued these penalties.[9] No further mention was made of the secret resolutions of the Confederation Congress regarding mail opening in the legislation passed under the Constitution.

Notes

1. Gaillard Hunt, ed., *Journals of the Continental Congress 1774-1789* (Washington, D.C: Government Printing Office, 1914), 23:670.
2. Ibid., 23:672.
3. John Jay to the President of Congress, Office of Foreign Affairs, 2 September 1785, in John C. Fitzpatrick, ed., *Journals of the Continental Congress, 1774-1788* (Washington, D.C.: Government Printing Office, 1933), 29: 679-680.
4. Bernard Porter, *Plots and Paranoia: A History of Political Espionage in Britain 1790-1988* (London: Unwin Hyman, 1989), 16-17.
5. Wednesday, 7 September 1785, entry in Fitzpatrick, Journals, 29:685.
6. Monday, 23 October 1786, in ibid., 31:909.
7. Wednesday, 7 September 1785, in ibid., 29:685 note.
8. Richard Peters, ed., *The Public Statutes At Large of the United States of America, 1789-March 3, 1845* (Boston: Little Brown & Co.: 1853), 1:70 for temporary act, and for permanent legislation, cf. ibid., 1 :236-237.
9. Ibid., 1:360-361.

Chapter 8

Department of Finance and Foreign Affairs Codes

In 1781 the American government took another major step towards national unity through the adoption of the Articles of Confederation. Though first proposed in July 1776, the Articles met stormy opposition from many states over the issues of taxation, the right of the federal government to dispose of public lands in the west, and the granting of one vote to each member state. Compromises smoothed over the differences, and in early 1781 Maryland finally ratified the Articles, a stronger government commenced, and new administrative offices were born.

The Department for Foreign Affairs was administered by a brilliant secretary, Robert R. Livingston, who would, for the first time, provide stability and enlightened leadership for this critical area. Like John Jay, Livingston was born in New York City in 1746, one year after Jay's birth, and completed his college studies at King's College. After several years of law studies, he was admitted to the bar in 1770, and joined a partnership with John Jay. He served three very active terms in the Continental Congress as a delegate from New York and served on numerous committees, including foreign affairs, military problems, and financial affairs.

Elected to the post, Livingston drafted, despite the poor communications across the Atlantic, the instructions for the American ministers at Paris, and helped to establish the basic principles for peace with Great Britain.

As Livingston emphasized, written communications between the American government and its ministers were basic components for the conduct of a rational foreign policy. A careful and thorough manager, he established procedures for an orderly processing of correspondence. From three to seven copies of each dispatch were prepared, and the original and three to five copies were carried by different ships across the Atlantic Ocean in the hope that at least one copy would reach its destination promptly and safely. In practical terms, promptly could be defined as approximately six weeks for a message from Philadelphia to arrive in Paris.

Livingston's frustrations about the lack of information arriving from America's ministers poured out in his dispatch to Benjamin Franklin in Paris in 1782: "It is commonly said that republics are better informed than monarchs of the state of their foreign affairs and that they insist upon a greater degree of vigilance and punctuality in their ministers." And with bitterness, Livingston continued, "We, on the contrary, seem to have adopted a new system. The ignorance in which we are kept of every interesting event renders it impossible for the sovereign to instruct their servant, and of course forms them into an independent privy council for the direction of their affairs, without their advice or concurrence." With barely controlled exasperation, the secretary added, "I can hardly express to you what I feel on this occasion. I blush when I meet a member of Congress who inquires into what is passing in Europe. When the General applies to me for advice on the same subject, which must regulate his movements, I am compelled to inform him that we have no intelligence but what he has seen in the papers. The following is an extract of his last letter to me: But how does it happen that all our information of what is transacting in Europe should come through indirect channels or from the enemy?"[1]

Exactly three months after Livingston penned his embittered dispatch, the elderly Franklin began his reply, his apologia, and explained the many obstacles to conducting a prompt and regular correspondence: distance from the seaports, promises that a ship would sail in a week or two and instead, in wartime, it would lie in port for months with dispatches aboard, waiting for a convoy. And he highlighted a fundamental security issue: "The post-office here is an unsafe conveyance; many of the letters we received by it have been opened, and doubtless the same happens to those we send; and at this time particularly, there is so violent a curiosity in all kinds of people to know something relating to the negotiations, and whether peace may be expected, or a continuance of war, that there are few private hands or travellers that we can trust with carrying our despatches to the sea-coast; and I imagine that they may sometime be opened and destroyed because they can not be well sealed."[2]

For a better perspective, Franklin added that European governments could receive reports from their ministers in Paris in ten to fifteen days, and indeed, answers could be gotten in that time. In conclusion, he added the hope that the American government would leave more to the discretion of her diplomats since five to six months were required before a reply to a query was received.

Franklin's explanations about the handicaps, particularly intercepted dispatches by foreign governments, served to renew Livingston's dedication to improve the use of better codes and ciphers.

Several months before Livingston came to the foreign office, a covey of new codes appeared: two-part codes developed for diplomatic correspondence, possibly by Robert Morris or Charles Thomson. The first official code under the Confederation was prepared on separate encode and decode sheets, with 660 printed numbers, containing 600 words, syllables, and letters of the alphabet scattered randomly throughout the decode sheet. Sixty blank numbers could be filled in later with a vocabulary particular to a minister's need in his country, or by a secretary or finance minister in Philadelphia.

This new code development eliminated a major weakness in the Tallmadge one-part alphabetical-numerical code, designed for espionage, by having two sheets, one in numerical order for decoding and the other in alphabetical order for encoding. Code-breakers would have many more variables to consider as they sought to decode intercepted dispatches.

These new foreign office codes had several key characteristics: a dot above a number represented an e ending for a word, and a dot below the number signified a plural ending. Twenty out of the twenty-seven most frequently written words in the English language were designated by a single number. Moreover, a conscientious code clerk (though they were rare) could write the word the by "278" with a dot above, or by a combination of numbers: "196" for t; "579" for h; "197" for e. Though these various methods were available, almost always, the would be encoded by "278."

The first recorded use of this new two-part system was by Lieutenant Colonel John Laurens, a daring soldier and the son of Henry Laurens of South Carolina. Sent on a special mission to France in 1781 (his father was then a prisoner of war in the Tower of London, having been captured at sea earlier), the colonel employed the code to veil his report on an interview with the French foreign minister, Count de Vergennes, during which Laurens underscored the desperate American needs for artillery, arms, tents, drugs, surgical instruments and military stores.

Robert Morris, the Philadelphia banker and merchant who became superintendent of finance, used a similarly designed code for his correspondence with John Jay, minister to Spain and urged him to request large loans or subsidies from the Spanish court. This sheet, with 660 printed numbers on the decode sheet, contained words, syllables, and letters of the alphabet carefully written down by Gouverneur Morris, assistant superintendent of finance. The encode sheet, with code elements printed in alphabetical order, listed the number after each element. The code was entitled "Office of Finance Cipher Number 1."[3]

A	456
ab	375
able	487
above	270
about	293
ac, ack, ak	296
act	434
ad	428
ag	181

Encode Sheet

The dire financial conditions in the nation prompted the increasingly anxious Morris to also write to Franklin in Paris. After relating the serious fiscal issues in his letter to Jay, Morris told him he hoped the French monarch would support the financial request made of Spain. He also hoped that the French government might consider a plan for refinancing the United States certificates. Though identical in format to the code he sent Jay, the elements were numbered differently, and it was labeled "Office of Finance Cipher No."[4]

Still another similar code of 660 elements, prepared by Secretary of Congress Charles Thomson, was sent to Jay from Livingston in July 1781, and modified somewhat a year later. The question why Livingston and Morris did not use the same code for Jay remains unanswered. Probably, Livingston believed a distinctly separate code offered greater security, and also enabled him to make modifications to his code, as he indeed did, without consulting Morris.

The most secure, complex, though awkward, codes used for diplomatic correspondence were 2,400-element forms with words and other alphabet clusters printed on an encode sheet (with numbers written on); and numbers from 1 to 1,000 printed on the decode sheet (elements written in). Robert Livingston sent such a code of 1,017 numbers from Philadelphia to George Washington on 27 June 1782, together with a letter of instructions that specified that when more than one word was represented by the same number, the correspondent was to draw two strokes under the second word and three strokes under the third. Livingston thought this would seldom be necessary, except occasionally at the beginning of a sentence before the sense was sufficiently plain to indicate the correct word.

This code and a similar one sent by Livingston to John Adams in The Hague (Adams did not receive it until May 1783, a year after it was sent!) and to Francis Dana in Russia (who never received it) were the most complex and when employed properly, promised the best security because they offered 2,400 elements. The most common among the twenty-seven most frequently used words were represented by two or more numbers in the Livingston-Washington code: for example, the was represented by "358," "447," and "507." This mask also contained a most significant element for confusing codebreakers, i.e., using a single number to signify a phrase. Thus, "401" represented "the U.S. in C. assembled."[5]

These creative new code forms in the early 1780s provided American ministers overseas with a more simple instrument than the Lovell ciphers, and a more efficient tool than the Dumas code. The problem remained, however, of delivering the code sheets safely to the correspondents, and then maintaining communications security through careful and restricted handling. Though couriers, such as Major David S. Franks, would be occasionally employed between 1781 and 1784, they were expensive, and budget-conscious managers were reluctant to send them except for major dispatches or treaties. Thus the security issues would continue to trouble U. S. government officials and their overseas ministers in the decades ahead.

Notes

1. Robert L. Livingston to Benjamin Franklin, Philadelphia, 2 September 1782, in Francis Wharton, ed., *The Revolutionary Diplomatic Correspondence of the United States* (Washington, D.C.: Government Printing Office, 1889), 5:696.

2. Benjamin Franklin to Robert Livingston, Passy, 5 December 1782 in ibid., 4:110-111.

3. Robert Morris to John Jay, Philadelphia, 7 July 1781, in ibid., 4:531-539. Jay used this code for sections of his lengthy dispatch to Livingston from Madrid on 28 April 1782: this message, containing more than 20,000 words, apparently was the longest dispatch sent to the Congress by an American minister during these years: cf. the dispatch in ibid., 5: 336-377.

4. Robert Morris to Benjamin Franklin, Philadelphia, 13 July 1781, in ibid., 4:571. This code was also written out by Gouverneur Morris.

5. Robert Livingston to George Washington, Philadelphia, 28 June 1782, in *Papers of George Washington*, Microcopy, Roll 86, Library of Congress.

Chapter 9
Jefferson-Patterson Ciphers

Soon after becoming president of the United States, Thomas Jefferson received a imaginative proposal from Robert Patterson for a novel cipher system, one that clearly surpassed the earlier and simpler designs of James Lovell. Patterson, an immigrant to America in 1768 from the north of Ireland, became a schoolmaster near Philadelphia for a short time before returning to that city and continuing his private studies. He established a country store in New Jersey several years before the Revolution, but after a year he closed that establishment and became principal of the Academy in Wilmington, Delaware.

After the outbreak of hostilities with Great Britain, Patterson, who had been a sergeant in a militia company in Ireland, became the drill instructor for the local militia and served as a brigade major for several years. In 1779, hc was appointed professor of mathematics and, a brief time later, became vice-provost at the newly established University of Pennsylvania. Elected a Fellow of the American Philosophical Society in 1783, he became secretary and in 1799, vice-president, shortly after Thomas Jefferson was chosen as president of the society. They served together until Jefferson's resignation in 1815. In 1805, Jefferson selected the highly intelligent mathematician as director of the United States Mint.

A very precise person, Patterson continually sought to transform and transfer abstract mathematical studies into practical applications. His books on mechanics, astronomy, and practical arithmetic introduced American students to both the theory and practice of numbers.

In a letter that Jefferson received on Christmas Day, 1801, the studious Patterson wrote at length from Philadelphia about a new cipher system. "A perfect cypher," he wrote, "should possess the following properties:

1. It should be equally adapted to all languages.

2. It should be easily learned & retained in memory.

3. It should be written and read with facility & dispatch.

4. (which is the most essential property) it should be absolutely inscrutable to all unacquainted with the particular key or secret for decyphering."[1]

Patterson fully believed his novel system readily met the first three conditions, and for the fourth, "it will be absolutely impossible, even for one perfectly acquainted with the general system, ever to decypher the writing of another without his key."

Continuing his instructions, Patterson wrote, "In this system, there is no substitution of one letter or character for another; but every word is to be written at large, in its proper alphabetical characters, as in common writing: only that there need be no use of capitals, pointing, nor spaces between words; since any piece of writing may be easily read without these distractions.

"The method is simply this — Let the writer rule on his paper as many pencil lines as will be sufficient to contain the whole writing. Then, instead of placing the letters one after the other as in common writing, let them be placed one under the other, in the

Chinese manner, namely, the first letter at the beginning of the first line, the second letter at the beginning of the second line, and so on, writing column after column, from left to right, till the whole is written.

"This writing is then to be distributed into sections of not more than nine lines in each section, and these are to be numbered (from top to bottom 1. 2. 3. &c 1. 2. 3.) The whole is then to be transcribed, section after section, taking the lines of each section in any order at pleasure, inserting at the beginning of each line respectively any number of arbitrary or insignificant letters, not exceeding nine; & also filling up the vacant spaces at the end of the lines with like letters.

"Now the key or secret for decyphering will consist in knowing the number of lines in each section, the order in which these are transcribed and the number of insignificant letters at the beginning of each line — all which may be briefly, and intelligibly expressed in figures, thus

"For example, let the following sentence be written in cypher according to the above key: Buonaparte has at last given peace to Europe! France is now at peace with all the world. Four treaties have been concluded with the chief Consul within three weeks, to wit, with Portugal, Britain, Russia, and Turkey. A copy of the latter, which was signed at Paris on Friday, we received last night, in the French Journals to the nineteenth. The news was announced, at the Theatres on the sixteenth, and next day by the firing of cannon, and other demonstrations of joy."

58
71
33
49
83
14
62
20

The first rank of figures expressing the number and order of the lines in each section, and the 2d rank, the number of arbitrary letters at the beginning of each respective line

First Draft

1 b i n l e i h t s h e e e n a e e a r
2 u v c l s t i h i e d c f i n s x n a
3 o e e t h h n p a l a e r n n o t n t
4 n n i h a t t o a a t i e e o n d o i
5 a p s e v h h r n t p v n t u t a n o
6 p e n w e e r t d t a e c e n h y a n
7 a a o o b c e u t e r d h e c e b n s
8 r c w r e h e g u r i l j n e s y d o
1 t e a l e i w a r w s a o t d i t o f
2 e t t d n e e l k h o s u h a x h t j
3 h o p f c f e b e i n t r t t t e h o
4 a e e o o c k r y c f n n h t e f e y
5 s u a u n o s i a h r i a e h e i r
6 a r c r c n t t c w i g l n e n r d
7 t o e t l s o a o a d h s e t t i e
8 l p w r u u w i p s a t t w h h n m
1 a e i e d l i n y s y i o s e a g o
2 s f t a e w t r o i w n t w a n o n
3 t r h t d i w u f g e t h a t d f s
4 g a a i w t i s t n r h e s r n c t

Transcribed in Cypher

w s a t a i s p a p s e v h h r n t p v n t u t a n o
e a a o o b c e u t e r d h e c e b n s b s a b d e p d n o
e h n o e e t h h n p a l a e r n n o t n t u t i o h
n e m e y e e s a n n i h a t t o a a t i e e o n d o i
r t l c r w r e h e g u r i l j n e s y d o t h d s e a r
s e e o b i n l e i h t s h e e e n a e e a r t a n r m
a r p e n w e e r t d t a e c e n h y a n o a b i
u v c l s t i h i e d c f i n s x n a h e n y t e n r f
s d t r o d i e s u a u n o s i a h r i a e h e i r p
s t o e t l s o a o a d h s e t t i e u a h r d e i u y
f t s h o p f c f e b e i n t r t t t e h o x e o y p u
p o r t e r s p i e e e o o c k r y c f n n h t e f e y o
t l r l p w r u u w i p s a t t w h h n m e n t
e r r e t e a l e i w a r w s a o t d i t o f n g e
w h a r c r c n t t c w i g l n e n r d h f o w s h
e t t d n e e l k h o s u h a x h t j o r u i y i
s a n t r h t d i w u f g e t h a t d f s l t m
a d t r o d i i e g a a i w t i s t n r h e s r n c t
n o n o a e i e d l i n y s y i o s e a g o d l l a n n
s f t a e w t r o i w n t w a n o n x y o u r e h

"It will be proper that the supplementary letters used at the beginning and end of the lines, should be nearly in the same relative proportion to each other in which they occur in the cypher itself, so that no clue may be afforded for distinguishing between them and the significant letters. The easiest way of reading the cypher will be, after numbering the lines according to the key, and cancelling the arbitrary letters at the beginning of the lines, to cut them apart, and with a bit of wafer, or the like, stick them on another piece of paper, one under the other, in the same order in which they were first written; for then it may be read downwards, with the utmost facility. On calculating the number of changes, and combinations, of which the above cypher is susceptible even supposing that neither the number of lines in a section, nor the number of arbitrary letters at the beginning of the lines, should ever exceed nine, it will be found to amount to upwards of *ninety millions of millions* (equal to the sum of all the changes on any number of quantities not exceeding nine, multiplied by the ninth power of nine) nearly equal to the number of seconds in three millions of years! Hence I presume the utter impossibility of decyphering will be readily acknowledged."[2] Patterson's system thrilled the imaginative President Jefferson, and though he was exceedingly busy with the domestic and foreign demands of his presidential office, he carefully wrote out the complicated design of the cipher,[3] then modified it and described it as follows:[4]

Method of Using Mr. Patterson's Cypher

1st. operation: In writing the original paper which is to be cyphered, use no capitals, write the letters disjoined, equidistant, and those of each line vertically under those of the one next above. This will be greatly facilitated, by using common black-lines, chequered by black-lines drawn vertically, so that you may place a letter between every two vertical black lines. The letters on your paper will thus be formed into vertical rows as distinct as the horizontal lines, divide

2nd. operation: To Cypher. Divide the vertical rows of the page into vertical columns of 9 letters or rows in breadth each, as far as the letters or rows of the line will hold out. The last will probably be a fractional part of a column. Number the vertical rows of each column from 1. to 9. in regular order. Then, on the paper to be sent to your correspondent, begin as many horizontal lines as there are vertical rows in your original, by writing in the beginning of each of every 9. horizontal lines as many insignificant letters from 1. to 9. as you please; not in regular order from 1. to 9. but interverting [*sic*] the order of the numbers arbitrarily. Suppose e.g. you 8. insignificant letters in the 1st line, 2 in the 2d. 1. in the 3d. 6 in the 4th. etc. you will thus have the horizontal lines of your 2d. paper formed into horizontal bands of 9. lines each, of which this, for instance, will be the key, or key of insignificant letters as it may be called. 8.3.1.6.9.4.7.2.5.1 2.9.1.8.4.6.3.7.5. / 3.6.9.2.8.5.7.4.1. / 2.1.3. Then copy the vertical lines of the 1st. paper, or original, horizontally, line for line, on the 2d. the columns in regular succession, put the vertical lines of each arbitrarily; as suppose you copy first the 1st. vertical lines of the 1st. column, the 5th. next, then the 2d. then the 8th. etc. according to this which may be called the key of lines 1.5.2.8.7.9.6.3.4. / 8.3.6.1.4.7.2.5.9. / 7.3.5.8.4.1.9.2.6. / 3.2.1. Then fill up the ends of the lines with insignificant letters, so as to make them appear of even lengths, & the work is done. Your correspondent is to be furnished with the keys thus:

key of letters 8.3.1.6.9.4.7.2.5. / 2.9.1.8.4.6.3.7.5. / 3.6.9.2.8.5.7.4.1. / 2.1.3.

key of lines 1.5.2.8.7.9.6.3.4. / 8.3.6.1.4.7.2.5.9. /7.3.5.8.4.1.9.2.6.13.2.1.

3rd. operation to decypher. your correspondent takes the cyphered paper you have sent him, & first, by the key of letters, he dashes his pen through all the insignificant letters, at the beginning of every line. Then he prefixes to the lines the numbers taken from the key of lines in the order in which they are arranged in the key. Then he copies the 1st. line of the 1st. horizontal band, writing on a separate paper, the letters vertically one under another (but no exactness is necessary as in the original operation.) he proceeds next to copy line No. 2. vertically also, placing it's [*sic*] letters by the side of those of his first vertical line: then No. 3 & so on to No. 9. of the 1st. horizontal band. Then he copies line No. 1. of the 2d. horizontal band, No. 2. No. 3. etc. in the regular order of the lines & bands. When he comes to the insignificant letters at the ends of the lines they will betray themselves at once by their incoherence, & he proceeds no further. This 3d. paper will then in it's letters and lines be the true counterpart of the 1st. or original.

Jefferson sent an enthusiastic appraisal of the cipher system to Patterson and described his modifications: "I have thoroughly considered your cypher, and find it is much more convenient in practice than my wheel cypher, that I am proposing it to the secretary of state for use in his office. I vary it in a slight circumstance only." Then Jefferson explained: "I write the lines in the original draught horizontally & not vertically, placing the letters of the different lines very exactly under each other. I do this for the convenience of the principal whose time is to be economised, tho' it increases the labor of a copying clerk. The copying clerk transcribes the vertical lines horizontally. The clerk of our correspondent restores them to their horizontal position ready for the reading of the principal."[5]

Flattered by Jefferson's praise, Patterson replied several weeks later and offered still another important modification: "There is yet another alteration, relative to the Key, which I conceive, would be of considerable advantage. Instead of expressing it by figures which are so liable to be forgotten, it may be expressed by a single word or name which may always be remembered, without committing it to writing." Then Patterson offered this example:

Suppose the key-word Montecello — the letters of this word are to be numbered according to their places in the alphabet, any letter repeated being referr'd to a second, or third alphabet – then the letters in the above word be numbered as follows

M	**o**	**n**	**t**	**e**	**c**	**e**	**l**	**l**	**o**
4.	**6.**	**5.**	**7.**	**2.**	**1.**	**8.**	**3.**	**9.**	**10**

the second e. l. and o. being referr'd to a second alphabet, and according [*sic*] numbered 8. 9. 10. This key-word will then signify that there are ten vertical lines in the section, which are to be transcribed in horizontal lines in the order of the above figures viz. 4th 6th 5th &c. The same word may also be used to signify the number of supplementary or insignificant letters at the beginning of the respective lines, as 4 at the beginning of the first, 6 at that of the second &c.[6]

Jefferson approved this latest design and wrote that this new cipher design would be used in foreign correspondence. However, he raised an important problem: "It often happens that we wish only to cypher 2 or 3 lines, or one line, or half a line, or a single word, it does not answer for this. Can your [cipher] remedy it."[7]

The innovative Patterson promptly replied and suggested that for enciphering a single word or line, the general system could be followed, and each letter should be considered as a column or a vertical line, and with supplementary letters prefixed and adjoining, should be transcribed into a horizontal line. Based on this pattern, he gave this example:

B	**e**	**n**	**j**	**a**	**m**	**i**	**n**		**F**	**r**	**a**	**n**	**k**	**l**	**i**	**n**
2	3	7	5	1	6	4	8		2	7	1	6	4	5	3	8

And then he explained that in order to encipher the plaintext word Louisiana, the design would look as follows:

2		**o**			
3				**u**	
7	**a**				
5			**s**		
1		**l**			
6			**i**		
4		**i**			
8					**n**
9		**a**			

Thus the order of the letters would be set by "Benjamin" and the number of nulls before the letters defined by "Franklin."

The plan for using key words to establish line placement and the number of nulls appealed to Jefferson, and three days after he received Patterson's 12 April letter, the president wrote a private letter to the American minister in Paris, Robert Livingston, and gave it to his friend, Monsieur Du Pont de Nemours, who was returning to Paris from Washington. Ever sensitive to the issue of maintaining secret codes and ciphers, Jefferson explained to his French friend, "The first page respects a cypher, as do the loose sheets folded with the letter. These are interesting to him and myself only, and therefore are not for your perusal."[8]

Jefferson promised Livingston that the innovative cipher would give him some difficulty; however, once understood, it would be "the easiest to use, the most indecypherable, and varied by a new key with the greatest facility of any one I have ever known." [9] The cipher keys chosen by Jefferson for Livingston incorporated Patterson's 12 April suggestions, although modified somewhat. Clearly, Jefferson liked and adopted the system of using key words because

> if we should happen to lose our key or be absent from it, it is so formed as to be kept in the memory and put upon paper at pleasure; being produced by writing our names & residences at full length, each of which containing 27 letters is divided into 3. parts of 9 letters each; and each of the 9. letters is then numbered according to the place it would hold if the 9 were arranged alphabetically, thus
>
> 651279843 923178546 3142 85 7 6 9
> ro bert r l i v ingst ono f c le r m on t
>
> 9476 18523 218965 7 34 7693 12458
> t h om a sjef f er so n ofm onti ce l lo

Although months had been spent in devising this special design, Livingston never employed it, nor did the secretary of state or President Jefferson. Instead, Livingston reported to Secretary of State James Madison in the 1,700-element printed code that he brought to Paris in 1801.[10] Thus, although Patterson and Jefferson formulated a superb cipher, the American ministers either failed to understand it fully or resented the extra hours required for masking the messages.

Notes

1. Robert Patterson to Thomas Jefferson, Philadelphia, 19 December 1801, in *The Papers of Thomas Jefferson*, 101 reels, Presidential Papers, Library of Congress, Roll 41.

2. Ibid.

3. According to an editor's notation in the original *Jefferson Papers*, Jefferson worked on the Patterson cipher on 12 April 1802, after receiving a Patterson letter on that date; however, internal evidence indicates that Jefferson worked out the system soon after the 19 December 1801, letter since he did not incorporate Patterson's idea of using a key word or name instead of figures that Patterson first introduced in his 12 April letter. Cf. *Jefferson Papers*, Roll 94.

4. Jefferson's Instructions in ibid., roll 44.

5. Jefferson to Patterson, Washington, D. C., 22 March 1802, in ibid., Roll 42.

6. Patterson to Jefferson, Philadelphia, 12 April 1802, in ibid., Roll 42.

7. Jefferson to Patterson, Washington, D.C., 17 April 1802, in ibid., Roll 42.

8. Jefferson to DuPont de Nemours, Washington, D.C., 25 April 1802, in Andrew A. Lipscomb, ed., *The Writings of Thomas Jefferson* (Washington, D.C.: The Thomas Jefferson Memorial Association of the United States, 1903-1904), 10:316.

9. Jefferson to Livingston, Washington, 18 April 1802, in *Papers of Thomas Jefferson*, Roll 42.

10. Cf. Despatches from *U. S. Ministers to France, 1789-1906*, Microcopy 34, Roll 10, National Archives, for his letters dated 12 March, 11, 13, 17 April, and nineteen others written during 1803 that included encoded messages.

Chapter 10

Jefferson's Cipher Cylinder

Sometime prior to 22 March 1802, the brilliant scholar president, Thomas Jefferson, designed a magnificent wheel cipher, a device absolutely extraordinary and imaginative.[1] In his letter to Robert Patterson, he wrote that Patterson's cipher, sent to him several months earlier, was so much more convenient to use than his wheel cipher that he was proposing to James Madison, his secretary of state, to employ it for the department's correspondence. Although Patterson's cipher system offered splendid security, it was extremely time-consuming to encipher and decipher. Jefferson's cipher cylinder promised more prompt and efficient enciphered communications, provided the devices could be safely delivered to the correspondents.

Thomas Jefferson

Apparently Jefferson did nothing further with his cylinder, and the design was rediscovered among his papers in the Library of Congress in 1922. Someone else invented a similar system, and in that same year the U.S. Army adopted an almost identical device.[2]

Later, other government agencies would also adopt it, particularly the U.S. Navy. The measurements and other design specifications for this truly unique device were recorded by Jefferson as follows:[3]

Turn a cylinder of white wood of about 2. Inches diameter & 6. or 8. I. long. bore through it's [*sic*] center a hole sufficient to receive [*sic*] an iron spindle or axis of 1/8 or 1/4 I. diam. divide the periphery into 26. equal parts (for the 26. letters of the alphabet) and, with a sharp point, draw parallel lines through all the points of division, from one end to the other of the cylinder, & trace those lines with ink to make them plain, then cut the cylinder crosswise into pieces of about 1/6 of an inch thick, they will resemble back-gammon man with plan sides. number each of the, as they are cut off, on one side, that they may be arrangeable [*sic*] in any order your please. on the periphery of each, and between the black lines, put all the letters of the alphabet, not in their established order, but jumbled, & without order, so that no two shall be alike. now string them in their numerical order on an iron axis, one end of which has a head, and the other a nut and screw; the use of which is to hold them firm in any given position when you chuse it. they are now ready for use, your correspondent having a similar cylinder, similarly arranged.

Suppose I have to cypher this phrase. 'your favor of the 22d. is received'. I turn the 1st. wheel till the letter y. presents itself. I turn the 2d. & place it's o. by the side of the y. of the 1st. wheel. I turn the 3d. & place it's u. by the side of the o. of the 2d. 4th. & place it's r. by the side of the u. of the 3d. 5th. & place it's f. by the side of the r. of the 4th. 6th & place it's a. by the side of the f. of the 5th. and so on till I have got all the words of the phrase arranged in one line, fix them with the screw, you will observe that the cylinder then presents 25. other lines of letters not in any regular series, but jumbled, & without order or meaning, copy any one of them in the letter to your correspondent. when he receives it, he takes his cylinder and arranges the wheels so as to present the same jumbled letters in the same order in one line. he then fixes them with his screw, and examines the other 25. lines and finds one of them presenting him these letters 'yourfavorofthe22isreceived.' which he writes down. as the others will be jumbled & have no meaning, he cannot mistake the true one intended, so proceed with every other portion of the letter. numbers had better be represented by letters with dots over them; as for instance by the 6. vowels & 4. liquids, because if the periphery were divided into 36. instead of 26. lines for the numerical, as well as alphabetical characters, it would increase the trouble of finding the letters on the wheels.

When the cylinder of wheels is fixed with the jumbled alphabets on their peripheries, by only changing the order of the wheels in the cylinder, an immense variety of different cyphers may be produced for different correspondents. for whatever be the number of wheels, if you take all the natural numbers from unit to that inclusive, & multiply them successively into one another, their product will be the number of different combinations of which the wheels are susceptible, and consequently of the different cyphers they may form for different correspondents, entirely unintelligible to each other. for though every one possesses the cylinder, and the alphabets similarly arranged on the wheels, yet if the order be inverted, but one line, similar through the whole cylinder, can be produced on any two of them. 2. letters can form only 2. different series, viz. a.b. and b.a. say 1 X 2 2 add a 3d. letter. then it may be inverted in each of these two series as lst.2d. or 3d. letter of the series, to wit c.a.b./ c.b.a. [Jefferson continues at the top of the next page].

a.c.b./ b.c.a.

a.b.c./ b.a.c.

consequently there will be 6 series = 2 X 3 or 1 X 2 X 3.

add a 4th. letter. as we have seen that 3. letters will make 6. different series, then the 4th. may be inserted in each of these 6. series, either as the lst.2d.3d. or 4th. letter of the series, consequently there will be 24. series.= 6X4=1X2X3X4X5X6.

dcab / cdab / cadb / cabd
dacb / adcb / acdb / acbd
dabc / adbc / abdc / abcd
dcba / cdba / cbda / cbad
dbca / bdca / bcda / bcad
dbac / bdac / badc / bacd

add a 5th. letter. as 4. give 24 series, the 5th. may be inserted in each of these as the lst.2d.3d.4th. or 5th. may be inserted in each of these as the lst.2d.3d.4th. or 5th. letter of the series, consequently there will be 120 = 24 X 5 = 1 X 2 X 3 X 4 X 5.

add a 6th. letter. as 5. give 120. series, the 6th. may be inserted in each of these as the lst.2d.3d.4th.Sth. or 6th. may be inserted in each of these as the lst.2d.3d.4th.5th. or 6th. letter of the series, consequently there will be 720.= 120 X 6 = 1 X 2 X 3 X 4 X 5 X 6. and so on to any number.

suppose the cylinder be 6.I. long (which probably will be a convenient length, as it may be spanned between the middle finger & the thumb of the left hand, while in use) it will contain 36. wheels, & the sum of it's combinations will be 1 X 2 X 3 X 4 X 5 X 6 X 7 X 8 X 9 X 10 X 11 X 12 X 13 X 14 X 15 X 16 X 17 X 18 X 19 X 20 X 21 X 22 X 23 X 24 X 25 X 26 X 27 X 28 X 29 X 30 X 31 X 32 X 33 X 34 X 35 X 36. = (4648 etc. to 42 places!!) a number of which 41.5705361 is the Logarithm of which the number is 372 with 39 chyers [zeros] added to it."

Without doubt, as David Kahn, the masterful historian of cryptography, states, Jefferson deserves the title of "Father of American Cryptography" because his design is so significant. The Jefferson wheel cipher was the most advanced cipher of its era. Indeed, this development elevates Jefferson to a height above Blaise de Vigenére and Girolamo Cardano, two brilliant architects in the design of secret writing.

Notes

1. Thomas Jefferson to Robert Patterson, Washington, D.C., 22 March 1802, in *The Papers of Thomas Jefferson* 101 reels, *Presidential Papers*, Library of Congress, Roll 42.

2. Kahn, *The Codebreakers*, 195.

3. *Jefferson Papers*, R 44. A rough draft may be also found in R 94.

Chapter 11

A Classic American Diplomatic Code

Familiar with the insecurity of written communications, especially during the American Revolution, imaginative Founding Fathers continued the practice of encoded writing in the closing months of that war and in the hectic postwar years when the enemy might be within the union as well as outside U.S. borders. Keenly aware of the weak American union, nervous about future threats from abroad, and anxious about jealousies among the states, certain significant diplomats such as Thomas Jefferson and James Monroe, together with James Madison and Edmund Randolph, all from Virginia, carried on an extensive correspondence, heavily sprinkled with encoded sentences and paragraphs, treating domestic and foreign issues.

A code of 1,700 numbers was devised in the early 1780s, and it became the standard design for most of diplomatic correspondence for the next eight decades. The first in the 1700 series was one by Edmund Randolph, handwritten, and probably mentioned by Madison in early 1783.[1]

Randolph, born near Williamsburg, Virginia, in 1753, and a graduate from the College of William and Mary, served briefly as aide-de-camp to General George Washington before returning to Virginia and becoming that state's attorney general, mayor of Williamsburg, and, later, delegate to the Continental Congress, governor, and, when Jefferson resigned as secretary of state, successor to that post in 1794.

This extensive Randolph code suffered from the dangerous weakness of listing words in alphabetical-numerical sequence, similar to the Tallmadge espionage code. Thus, the codenumbers from "1" through "79" began with b, and ranged in alphabetical order from b through by before going to the D category. Similarly, codenumbers "201" through "300" represented A though ay. Other such lengthy sequences made the code much more vulnerable to codebreakers.

The encoded messages passing between these Virginia statesmen at this time included personal affairs such as Madison's courtship with Catherine Floyd,[2] the tense foreign affairs issues along with the petty personal rivalries being considered in the Congress,[3] and discussions on raising a general impost to restore the public credit of the United States.[4]

In the years following 1783, several other 1,600- and 1,700-item codes appeared. Among the most notable was a code sometimes referred to as "Jefferson's Third Cypher." Jefferson sent it from his post as U.S. minister to France under the care of young John Quincy Adams to Madison. Jefferson also sent a second identical copy to James Monroe so the three close friends and colleagues had a common instrument for classified correspondence. Curiously, Jefferson sent John Jay, then secretary of foreign affairs, a code different from the one he sent his two Virginia associates, perhaps to maintain a special correspondence with Madison and Monroe. Always anxious about maintaining the secrecy of their correspondence, Madison and Jefferson exchanged several other 1,700-item codes, especially while the latter was stationed in France.

As the new American government under the Constitution commenced in 1789 and Thomas Jefferson became secretary of state, foreign correspondence to and from Europe required the utmost secrecy, especially because of the constant intrigue along the U. S. borders and, as well, foreign trade competition confronting the new republic. During the 1790s, over 5,000 lines of encoded dispatches were sent by American ministers in France, Great Britain, Spain and the Netherlands. Over 1,800 lines alone were sent from the American commissioners, John Marshall, Charles C. Pinckney, and Elbridge Gerry, in France during the time of the XYZ Affair in 1797.

During the so-called XYZ Affair, gifts and loans from America were sought by Talleyrand and other agents of the French Directory; the Americans remained adamantly opposed. One of the encoded dispatches to the secretary of state, Timothy Pickering, reported the discussion about Jean Conrad Hottinguer (Mr. X) and his request for funds: "M. Hottinguer again returned to the subject of money: said he, gentlemen you do not speak to the point; it is money; it is expected that you will offer money: we said we had spoken to that point very explicitly: we had given an answer. No, said he, you have not: what is your answer? We replied, it is no, no, not a six pence."[5] Thus one of the more famous phrases in American diplomatic history was originally reported in code as

no	no	not	a	six	pen	ce
449.	449.	457.	1193.	1178.	27.	493.

In 1803, a new code of 1,700 items, which later came to be termed the "Monroe Cypher," accompanied James Monroe as he travelled to France as envoy extraordinary and minister plenipotentiary at the express request of President Thomas Jefferson. Earlier governor of Virginia and minister to France, Monroe joined Robert L. Livingston during negotiations with France for the purchase of New Orleans and lands to the east of the Mississippi River. The two Americans were successful in buying the Louisiana Territory.

First introduced in 1803 to mask the tense messages to and from the two American ministers in Paris, this code became the classic secret instrument for American ministers in Europe during the next six decades. Although a few other codes and ciphers would also be used during these years, the Monroe Cypher remained the standard tool for masking dispatches to other ministers and also to the State Department from Europe and Mexico.

rather	727
ration	728
rch	729
rd	730
re	731
rca	732

Encode section of the Monroe Code, 1803-1867

727	rather
728	ration
729	rch
730	rd
731	re
732	rca

Decode section of the Monroe Code, 1803-1867

Notes

1. James Madison to Edmund Randolph, Philadelphia, 18 March 1783, in William T. Hutchinson and William M. Rachal, eds., *Papers of James Madison* (Chicago: University of Chicago Press, 1969), 6:356.

2. Thomas Jefferson to James Madison, Susquehanna, 14 April 1783, in ibid., 6:459. Also cf. Irving Brant, *James Madison: The Nationalist, 1780-1787* (New York: Bobbs-Merrill, 1948) 2: 283-237 where the author notes the romance never blossomed.

3. James Madison to Thomas Jefferson, Philadelphia, 6 May 1783, in ibid., 7:18-19.

4. Ibid., Orange, 10 December 1783, 7:401-403.

5. Charles Pinckney, John Marshall, and Elbridge Gerry to Timothy Pickering, Paris, 8 November 1797, in *Dispatches from United States Ministers to France, 1789-1906,* 128 rolls, Record Group 59, Microcopy 34, Roll 8.

Chapter 12

John Quincy Adams's Sliding Cipher

In December 1798, a most unique and innovative sliding cipher, devised by John Quincy Adams, then minister in Berlin, was sent by its inventor to William Vans Murray, American ambassador to The Hague. One year earlier, the XYZ Affair and further misunderstandings regarding American support for the French leaders in their struggle against the British armed forces had severed diplomatic relations between the United States and France. And soon America fought an undeclared naval war against her former ally, France. Because of President John Adams's determined crusade for peace, and also due to his skillful dealings with a war-hungry Congress, this armed struggle did not erupt into a full-scale war. The Hague, during these years, became a crucial window for Murray who, in concert with the French envoy located there, arranged for further negotiations and reconciliation with France.

John Quincy Adams

Both Murray and Adams recognized the necessity for better diplomatic communications security, especially in war-ravaged Western Europe. Two decades earlier, Adams, then ten years old, traveled with his father to France in 1778, studying there and at Amsterdam and Leyden University before accompanying Francis Dana, the new American minister to Russia in 1781. Less than two years later, he returned to The Hague for further studies before going to Paris with his father for final peace negotiations with Great Britain. Returning to America, he completed his studies at Harvard College, studied law, and finding the profession unattractive, he turned to politics and soon won a commission in 1794 from President George Washington as minister to the Netherlands. However, French armies occupied that nation so Adams visited other major countries, particularly England, before being assigned to Berlin. Adams's foreign travels and first-hand experiences with espionage in France and Russia quickly taught him the critical skills for secret secure communication. In addition, both Adams and Murray were well aware that their dispatches could be used by one American political party or the other or, indeed, be published in the American newspapers.

Adams's deep anxieties about European intercept practices regarding foreign dispatches were accurate. During 1798 and 1799, British postal authorities and other agents seized almost thirty highly confidential dispatches from Murray to Rufus King, American minister to Great Britain; King to John Quincy Adams; Adams to King; and King to Murray. Most of the letters sent in cipher were broken by the British; however, only about 15 percent of those in code were read.[1]

John Quincy Adams's sliding cipher

	101	a					
	102	b					
	103	c					
	104	d					
	105	e					
	106	f					
	107	g					
	108	h					
	109	i					
	110	k					
	111	l					
	112	m					
	113	n					
	114	o					
	115	p					
	116	q					
	117	r					
	118	s					
	119	t					
	120	u					
	121	w					
	122	x					
	123	y					
	124	z					
	125	&					
Africa	126	A	1	2	3	4	London
America	127	B	5	6	7	8	Madrid
Against	128	C	9	10	11	12	Mediterranean
Army	129	D	13	14	15	16	Men
Article	130	E	17	18	19	20	Minister
Asia	131	F	21	22	23	24	Naples
At	132	G	25	26	27	28	Nation
Atlantic	133	H	29	30	31	32	Navy
Austria	134	I	33	34	35	36	Of
Baltic	135	K	37	38	39	40	Paris
Batavia	136	L	41	42	43	44	Parliament
Because	137	M	45	46	47	48	Peace
Berlin	138	N	49	50	51	52	People
But	139	O	53	54	55	56	Petersburg
By	140	P	57	58	59	60	Philadelphia
Cabinet	141	Q	61	62	63	64	Poland
Congress	142	It	65	66	67	68	Portugal
Constantinople	143	S	69	70	71	72	Power
Constitution	144	T	73	74	75	76	President
Copenhagen	145	U	77	78	79	80	Prussia
Council of 500	146	W	81	82	83	84	Republic
Council of antients	147	X	85	86	87	88	Rome
Country	148	Y	89	90	91	92	Russia
Court	149	Z	93	94	95	96	Sardinia
Denmark	150	&	97	98	99	100	Senate
Department	151	a	1	2	3	4	Sicily
Directory	152	b	5	6	7	8	Spain
East Indies	153	c	9	10	11	12	State
Egypt	154	d	13	14	15	16	Stipulation
Emperor	155	a	17	18	19	20	Stockholm
Empire	156	f	21	22	23	24	Sweden
Europe	157	g	25	26	27	28	Switzerland
Fleet	158	h	29	30	31	32	That
For	159	i	33	34	35	36	The
France	160	k	37	38	39	40	Then
From	161	l	41	42	43	44	There
General	162	m	45	46	47	48	This
Germany	163	n	49	50	51	52	Through
Government	164	o	53	54	55	56	To
Great Britain	165	p	57	58	59	60	Treaty
Hague	166	q	61	62	63	64	Turkey
House of Representatives	167	r	65	66	67	68	Tuscany
In	168	s	69	70	71	72	United States
Into	169	t	73	74	75	76	Upon
Ireland	170	u	77	78	79	80	Vienna
Italy	171	w	81	82	83	84	War
King	172	x	85	86	87	88	West Indies
Kingdom	173	y	89	90	91	92	Whether
Law	174	z	93	94	95	96	Which
Lisbon	175	&	97	98	99	100	With

Source: The Adams Papers, Massachusetts Historical Society
Note: In this code the center sliding strip was aligned according to the sender's instructions.

Though seven years older, Murray had much less experience than Adams with foreign cultures and surveillance: he did pursue three years of law studies in England soon after the American Revolution. Adams would become his instructor in secret writing. Three days before Christmas in 1798, Adams sent a novel cipher strip to Murray, and he carefully explained that the device must be kept confidential. Extremely cautious, he reported that instructions for the system would follow in another dispatch.[2] And on Christmas Day, Adams sent the second part of what he termed "my hieroglyphics." The precise written explanations told how to fit the sliding strip into the cipher sheet and instructed Murray to use the cipher only for corresponding with Adams. Also included with the explanations was a four-sentence paragraph in cipher for Murray to decipher.[3]

Murray, the eager pupil, in his dispatch on New Year's Day, thanked his friend for the "C," as he termed the cipher, which he had received just the previous evening "and can not for my life make it out. I have turned the affair upside down and downside down, this side and the other, placed the slip in all possible bearings, and 'have worked all night but caught no fish' – for I was at it till late" and pleading tired eyes, Murray turned to other less frustrating and complicated matters.[4]

In fairness to Murray, it must be noted that the device required some ingenuity, particularly because Adams's original instructions on Christmas Day simply said, "The first figure 21 is only to give the key — draw your strip so that the letter A shall stand opposite to that figure & you will then be able to decypher the whole." This brief instruction befuddled Murray much like James Lovell's cipher explanations to John Adams almost two decades earlier. Adams followed up with a more lengthy narrative one week later and was pleased to learn from his pupil that he had acquired the skills necessary for using the cipher. And during the next ten months, the correspondents composed confidential enciphered messages about the economic affairs of the Court of Vienna in relation to London; the determination of the empress of Russia to send 8,000 men into Italy; that the emperor of Germany left his own father-in-law to inevitable destruction, as the king of Prussia abandoned the House of Orange; and French spying.[5]

Several times during 1799, Adams's imaginative cipher still caused Murray difficulties. This brought Adams to comment, "My poor cipher! I meant to make it complicated & increase the difficulties of decyphering. And Lo! I made it unintelligible to my own correspondent.... You laughed at me for my great A, & little a, & something (bouncing) 'B' but my object in using a great & a small alphabet was, that the same letters might without repetition designate different words."[6] Embarrassed with his mistakes and forgetfulness, Murray hastened to praise Adams's creation, calling it the finest he had ever seen. And he added, he hoped the secretary of state would adopt the system and make up three or four different sets, for he feared the Adams-Murray model might be known: "But be assured also that the office itself is under the guard but of honesty, and has few barriers against the thousand ways and means by which papers are obtained in Europe."[7]

European espionage agents and practices continued to challenge American diplomatic ministers, conscientious about making their communications safe and secure. The John Quincy Adams sliding cipher provided a most fascinating though very limited response to these threats.

	SPAIN	FRANCE	GREAT BRITAIN	THE HAGUE	PRUSSIA	MEXICO	RUSSIA
1816	132	105	—	—	—	—	7
1817	—	13	—	—	—	—	—
1818	—	293	102	—	—	—	74
1819	—	56	25	—	—	—	—
1820	18	4	—	—	—	—	—
1821	—	8	22	—	—	—	38
1822	103	12	—	—	—	—	—
1823	12	18	—	—	—	—	—
1824	3	—	—	—	—	—	—
1825	140	—	—	—	—	707	—
1826	—	—	—	—	—	108	—
1827	—	—	10	—	—	—	—
1828	—	—	—	—	—	14	—
1829	—	—	—	—	—	50	—
1830	—	—	—	—	—	48	—
1831	—	—	—	—	—	32	—
1832	—	—	—	—	—	—	—
1833	—	—	—	—	—	11	—
1834	—	—	—	—	—	9	—
1835	—	—	—	—	—	—	—
1836	—	—	—	—	—	857	19
1837	—	—	—	—	—	—	—
1838	—	—	—	—	—	—	—
1839	—	—	—	—	—	86	—
1840	—	—	—	—	—	—	—
1841	144	—	—	—	—	—	—
1842	—	—	—	—	—	—	—
1843	—	—	—	—	—	—	—
1844	—	—	—	—	—	—	—
1845	—	—	—	—	—	—	—
1846	—	—	—	—	—	—	—
1847	—	—	—	—	—	175	—
1848	14	—	—	—	—	—	—

Source: Ralph E. Weber's *United States Diplomatic Codes and Ciphers, 1775-1938*

Encoded lines in American dispatches from European legations, 1816-1848

Notes

1. "Deciphers of Diplomatic Paper," LI, American 1780-1 841, British Museum, Add. Mss. 32303. These papers were reproduced for the Division of Manuscripts of the Library of Congress by Grace Gardner Griffin. Also cf. Samuel Flagg Bemis, "British Secret Service," *American Historical Review*, 29 (April 1924), 483.

2. John Quincy Adams to William Vans Murray, n.p., 22 December 1798, in the *Adams Family Papers*, 608 rolls. Massachusetts Historical Society, 1954-1956, Roll 133.

3. Ibid., 25 December 1798, in ibid., Roll 133.

4. William Vans Murray to John Quincy Adams, n.p., 1 January 1799, in Worthington Ford, ed., "The Letters of William Vans Murray," *Annual Report of the American Historical Association for the Year 1912,* 502.

5. Adams-Murray correspondence during January, May, September and November 1799, in *Adams Family Papers*, Roll 133,394,395,396, and Ford, "LWVM," 595.

6. John Quincy Adams to William Vans Murray, 2 November 1799, in *Adams Family Papers*, Roll 134.

7. William Vans Murray to John Quincy Adams, n.p., 12 November 1799, in Ford, "LWVM," 617; the original letter is in *Adams Family Papers*, Reel 396. At the top of the dispatch there was the notation, "This is almost too bad to trust by post – observe the seals."

Chapter 13

Aaron Burr's "Cipher Letter"

An impulsive adventurer, Aaron Burr, then fifty years old and living in Philadelphia, and still eager to acquire an empire, prepared a secret dispatch that was written in code and cipher and addressed to General James Wilkinson in July 1806. This was the famous letter that would lead President Thomas Jefferson to charge his former vice president and political enemy, Burr, with treason. On the basis of this dispatch, the cunning Wilkinson, who had become a secret, tenured, and well-paid informer ("Agent 13") of the Spanish government in the 1780s, warned Jefferson that a plot, led by Burr and described in the cipher letter, was under way to disrupt the United States and to take over Spain's colony, Mexico. Incidentally, Wilkinson's agent status, though long suspected, was finally documented by historians from records in the Spanish archives over a century later.

Wilkinson, with dreams of empire-building much like Burr, decided against a partnership in this enterprise, apparently designed by the two of them earlier. The troubled general betrayed his friend and collaborator likely in order to regain favor with Jefferson, who had just removed the unpopular Wilkinson from the governorship of the Louisiana Territory in May. Wilkinson's warning to Jefferson also served as evidence of his personal loyalty to the president and the United States, and indeed to Spain, whose North American empire was threatened.

Profound mystery surrounds the two copies of Burr's enciphered dispatch since Wilkinson admittedly falsified the original message, possibly changing passages that could be used as an indictment of Wilkinson. The general sent Jefferson his own plain text of Burr's letter, never producing the actual enciphered dispatch. The original dispatches in Burr's writing have never been discovered. Historians and biographers of Burr have given different decipherments of that enigmatic July 1806 letter: some of the sentences were based upon a book, possibly *Entick's Pocket Dictionary*, or Pron't [sic] *English Dictionary*, with page numbers and columns masked by unknown additives; moreover, the place and date of publication of the particular dictionary remain uncertain.

Like many of his political contemporaries such as Jefferson, Madison, and Monroe, and eager to hide his confidential letters from curious eyes, Burr occasionally turned to secret writing. In February 1801, he drafted a note in code to Congressman Edward Livingston and included a political code of 180 items, largely in numerical order, that provided Arabic numbers for the names of congressmen, states, cities, federal cabinet offices, and officers.[1] In several instances, in an attempt to increase security, he assigned two different numbers for the same person: for example, Jefferson was "127" and "128" and Burr, "129" and "130."

On 22 July 1806, Burr had prepared two copies of his famous letter for General Wilkinson, a friend since their military days in the American Revolution. His lengthy relationship with the unscrupulous Wilkinson reflected his poor judgment of individuals. As Andrew Jackson remarked, "Burr is as far from a fool as I ever saw, and yet he is as easily fooled as any man I ever knew."[2] One copy of Burr's letter was carried by a loyal supporter, Samuel Swartwout, who probably wrote the cipher in the original Burr letter, and a second copy was carried by another associate, Dr. Justus Erich Bollman.

Each copy must have required three to five hours to encipher and encode exactly. Both copies reached Wilkinson, the former in October and the latter in November. The most recent biographer and defender of Burr argues that Wilkinson never received Burr's original letter: "The cipher letter as we know it today was not written by Aaron Burr. It was written by Jonathan Dayton."[3] He contends that Dayton, a former senator from New Jersey, and collaborator with Burr, prepared two new letters and substituted them secretly for the ones carried by Swartwout and Bollman. No copy of the original letter in Burr's handwriting has been found, and Burr declared the dispatch provided to Jefferson by Wilkinson a forgery.

According to this fascinating account in the recent Burr biography by Milton Lomask, Dayton's copy included some of Burr's original sentences but included more flamboyant phrases about the exciting opportunities and strong auxiliary support for winning the new empire. Dayton hoped to increase Wilkinson's enthusiasm and backing for the expedition. Lomask conjectures that Dayton destroyed the authentic Burr letter.

Preparing the Burr original dispatch in code and cipher required the accuracy of a certified public accountant and the patience of a medieval monk. The secret system combined three complicated patterns of secret writing. One form, as noted above, was a book code, with a page number and word position in the column designated by two numbers, probably with the page number disguised by a specified additive and a line over the number referring to one of two columns on the page; the second number indicated the position of the word on the page. Secondly, Arabic numerals served as masks with "13," "14," "15," and "16" for Burr, and "45" for Wilkinson. Third, over sixty-six various symbols were used to represent letters of the alphabet and also various words. For example, a short horizontal line was a and a short vertical line, b; a square represented England, a square with a dot in the middle, France; a circle, President, and a circle with a dot in the middle, Vice President.[4]

Briefly, the mysterious two-page Burr dispatch of July 1806, as deciphered by the editors of the Burr Papers, announced that funds for the expedition had been obtained and that six months provisions would be sent to locations named by Wilkinson. Further, groups would gather on the Ohio on 1 November and move to Natchez, meeting Wilkinson there in early December. "The people of the country to which we are go[ingl are prepared to receive us — their agents, now with me, say that if we will protect their religion and will not subject them to a foreign power, that in three weeks all will be settled. The god invite us to glory and fortune. It remains to be seen whether we deserve the boons."[5]

Following Wilkinson's betrayal, Burr was tried in the spring and summer of 1807 in Virginia for a misdemeanor for organizing an expedition against a foreign colony, Mexico. Based on Wilkinson's accusations, the charge of treason was added by a grand jury, basing its decision upon a mistaken understanding of Chief Justice John Marshall's earlier obiter dicta in the trials of Burr's messengers, Swartwout and Bollman. A jury acquitted Burr of treason in September when it was shown that he had not actually participated in an overt act of levying war. The jury also ruled against the misdemeanor charge; however, Burr was to be sent back to the district court in Ohio. He never returned, and the government failed to press the law suit.

In July 1808, Burr sailed for England, and after several months visited Scotland before being expelled from Great Britain. Probably while in England, he prepared another code, this one for his correspondence with his only child, Theodosia. This curious mask would disguise the months and days and used almost fifty Arabic numbers, numerous pronouns, verbs, and amounts.[6] A lengthy stay in Sweden and extended travel through Germany, France, Holland, and a return to England during the next three years, preceded his return to the United States a month before the War of 1812 with England broke out. Resuming the practice of law and his fascination with profitable business opportunities in the Latin American colonies and the later republics, Burr continued his energetic quest for adventure and economic power until his death on Staten Island in 1836.

Notes

1. Mary-Jo Kline, ed., *The Papers of Aaron Burr, 1756-1836* (New York: Microfilming Corporation of America, 1978), Roll 4,876-77.

2. Milton Lomask, *Aaron Burr, The Conspiracy and Years of Exile, 1805-1836* (New York: Farrar, Straus, Giroux, 1982), 13.

3. Ibid., 118.

4. Kline, Papers, Rolls 6, 166, provides the key to the various symbols used by Burr; however, the book code explanation is not included. The editors of the Papers compared the different cipher copies owned by the Newberry Library with other "accepted" versions and reconstructed the dispatch, especially those sentences based on the book code. This version may be found on Rolls 6, 170. Moreover, this decipherment differs from copies in the Annals of Congress, 9th Congress, 2d Session, 1011-1012, and 1013-1014, which were reprinted from General Wilkinson's dispatches to Jefferson of December 1806.

5. Ibid., Roll 6, 170.

6. Ibid., Roll 6, 498.

Chapter 14

The First U.S. Government Manual on Cryptography

David W. Gaddy

It is always risky to claim a "first," especially in such an esoteric field as cryptography, and government cryptography at that. But mounting evidence points to a small book with the simple title *Cipher* as a prime candidate for the U.S. government's first manual on the making and breaking of ciphers, predating by a half century or more the army manual by Parker Hitt that has been accorded that position. Carrying no explicit evidence of the publisher, date, or place of publication, the book is a reprint of a lengthy encyclopedia article on the subject penned by an English surgeon and amateur student of cryptography, William Blair, around 1807. It exists in two editions, a 117-page version (presumed to be the earlier) and an expanded, 156-page version, which alludes to practices during the American Civil War. The book was evidently printed by and for the Chief Signal Officer of the U.S. Army in the mid-nineteenth century and was used as a text in the instruction of signal officers.

Around 1807, Dr. William Blair (1766–1822) wrote for Abraham Rees's *Cyclopaedia* a lengthy article (some 35,000 words) under the heading of "Cipher."[1] According to internal evidence, the forty-one-year-old surgeon was an amateur drawn to the subject by a museum exhibit that kindled his interest. For three years he must have read everything he could get his hands on, in English, French, Latin, Greek, and possibly other languages. Dissatisfaction with the treatment of the subject in available authorities (such as *Britannica*) led him to undertake a superior presentation, and in that he ably succeeded. David Kahn, in his monumental work *The Codebreakers* (1967), characterized Blair's "superb article" as "the finest treatise in English on cryptology" until Parker Hitt's military manual was published by the U.S. Army in 1916.[2] Rees issued his work in parts, comprising thirty-nine volumes, over a period of eighteen years, the Blair article appearing in May 1807. An American edition followed on the the heels of the London edition.[3] One reader who evidently digested its contents was Edgar Allan Poe, popularly regarded as a cipher expert because of "The Gold Bug" and other writings in the late 1830s/1840s.[4] Yet, while American Civil War contemporaries of the 1860s cited "The Gold Bug" as the basis for their familiarity with letter-frequency in cipher solution,[5] Rees's half-century-old encyclopedia seemed to have fallen out of use and common knowledge by that time. Ironically, Blair (citing Falconer and other authorities) described the principal cryptosystems that would be used by the two sides in the American Civil War – the Confederacy's dictionary cipher, the Vigenère (as it is generally known today) and the grille, as well as simple substitution ciphers, and the Union's word-route transposition cipher - and suggested ways of solving them. In the last case, that of the route transposition – perhaps the earliest American cipher designed specifically for the telegraph - contemporaries ascribed its origin to Anson Stager, creator of the U.S. Military Telegraph Department (later "Corps"), evidently never having heard of the Earl of Argyll (1685), James Falconer (1685), or Blair.[6]

With respect to the Civil War period, the first Signal Officer of the U.S. Army (1860), Albert James Myer (1828-1880), inventor of the "wig-wag" system of visual signaling, was "bounced" from his position as Chief Signal Officer in 1863, in the middle of the war, and was not restored until 1867, two years after the fighting ended.[7] In 1864 Myer had hastily printed a signal manual to which he appended, as a section on ciphers, an article drawn from a popular journal of 1863. After the war, he enlarged on the provisional text, producing the first in a series of copyrighted editions that made up the principal manual and reference ("Myer's Manual") for military communicators for the remainder of the nineteenth century, but he retained the 1863 cipher article. Like Blair, Myer, the "Father of the Signal Corps" (and Weather Bureau, among other accomplishments), was also a surgeon, with a scientific background. He pushed his people into the study of electricity and electromagnetic telegraphy, into balloons and aeronautics, into meteorology, and into telephony. His appreciation of signal or communications security stemmed from wartime experience with an able adversary who had jumped the gun on him in his chosen field. But he seemed to lack anything better on the subject of cryptography than a popular journal article. Perhaps that is an incorrect impression.

The small book that constitutes our subject appears in David Shulman's *An Annotated Bibliography of Cryptography* (1976), with the authorship correctly interpolated (William Blair's name had been used as a example in the text), as was the fact that it is a reprint of the Rees article. He notes a copy of the 117-page version in his personal collection (subsequently placed in the New York Public Library), plus one in the Harvard University Library, and another, enlarged edition (156pp., vice 117) at the latter place. Inquiry in 1990 disclosed that Shulman was in error in listing Harvard and that the Boston Public Library was the actual repository. According to the Rare Books staff in Boston, their copy of the earlier edition has been missing since 1936, but their other copy – the expanded edition – bears the identification of "Philip Reade, 2nd Lt, 3rd Inf, Actg Signal Officer, San Diego, 1875."

Recently a private collector has reported the acquisition of a copy of the expanded version that clarifies some aspects of this book, while raising other questions. It bears a handwritten inscription on the flyleaf as follows:

> **This little volume is of interest, being the [?textbook?] of the Signal Officer class of 1869, of which I was, fortunately, the only officer to pass successfully the final examination.**
>
> **/s/A.W. Greely**
>
> **Lieut 36th US Inf**
>
> **Acting Signal Officer**[8]

In 1989 a copy of the shorter (presumed earlier) edition was donated to NSA. Curiosity prompted this study. The book is marked in red ink on the front fly leaf "Office Copy" and, both on the title page and first page, "Signal Department USA No. 18." It has been hand corrected in pencil where typographic errors appear in the text, showing careful study (but overlooking a typo misspelling "Vigenère" as "Vigeuere" in one place). There is a copy of this same edition, privately re-bound, and with no marking, in the William F. and Elizebeth S. Friedman Collection at the George C. Marshall Research Library in Lexington, Virginia, without elucidation in the card catalog beyond the fact that it was a reprint of Blair, suggesting that Mr. Friedman appreciated it only as a historical curiosity. The Library of Congress also holds a copy of this edition, transferred there from the "W. B. [Weather Bureau?] Library" in 1917 and re-bound by the Library of Congress.[9]

The difference between the two editions is that the smaller one is a verbatim, type-reset and paginated reprinting of the 1807

Blair article from the Rees encyclopedia or its American counterpart. The expanded version (to be dated at least by 1869) appends some later information and observations based upon experience during the American Civil War. A record of the course of instruction for the 1869 signal officers' class notes the use of the Myer manual and a cipher manual, which would accord with Greely's notation.[10] But if Greely is to be taken literally, that the expanded version in which he wrote was the one used in his class just four years after the war ended, and if our speculation is correct that the shorter version predated it, we are left wondering ... by how many years? The cryptologic state of knowledge of Civil War participants does not appear to be consistent with the comprehensive basis afforded by Blair. Further research in the correspondence and contracting records of the Chief Signal Officer at the National Archives might fix dates and specifics and thereby shed more light on a dim chapter of early American cryptography.

Having continued in postwar editions of his *Manual* to use the 1863 journal article as if he knew of no superior treatment of cryptography, the Chief Signal Officer, one might speculate, subsequently learned of the Blair piece and had it printed to offer his men a concise supplemental reference on cryptology, much as a modern instructor might refer his students to the classic encyclopedia articles by William F. Friedman or Lambros D. Callimahos. Modern practices would expect identification of the source and authorship, however, and in this case, neither was given (if known). Perhaps this was considered not plagiarism, but a "lifting" for limited "official use only," thereby accounting also for the evidently small number of surviving copies."[11] In any event, Myer's use of "the finest treatment of the subject in English" remains a tribute to the professionalism he sought to instill in the Signal Corps and affords new insight into American military knowledge of cryptography during what had been regarded as the cryptographic "dark ages" of the mid-nineteenth century. And, in turn, one is left wondering. . . was Parker Hitt aware of Blair and "the little black book," or was there yet another gap in institutional memory?

Notes

1. An excellent overview of the work of Dr. Abraham Rees (1743-1825) is contained in Harold R. Pestana, "Rees's Cyclopaedia (1802-1820), A Sourcebook for the History of Geology," in the *Journal of the Society for the Bibliography of Natural History*, Vol. 9, No. 3 (1979), 353-61. The Cyclopaedia was issued serially in eighty-five parts over two decades, two parts intended to comprise a volume (except for XXXIX, which required three parts), plus plates. Individual pages are not numbered. From extant library copies, some subscribers did not take the trouble to bind the parts as intended. That, plus the fact that individual articles are not signed, might well have caused a casual reader to overlook the fact that William Blair was identified in the preface (v) as the author of the article on "Cipher," but a careful reader would have discovered his identity buried in the text.

2. Kahn, *The Codebreakers*, 788, echoing the *British Dictionary of National Biography*, "incomparably the best treatise in the English language on secret writing and the art of deciphering" (V, 168). Parker Hitt, *Manual for the Solution of Military Ciphers* (Fort Leavenworth, Kansas: Press of the Army Service Schools, 1916, First Edition).

3. Pestana, "Rees's Cyclopaedia," 356-357. The American edition appears to have commenced in 1806, expanded to include items considered to be of specifically American interest. Robert A. Gross, *Books and Libraries in Thoreau's Concord* (American Antiquarian Society, 1988) lists in that library (415) Rees's "First American Edition, Revised, Corrected, Enlarged, and Adapted to this Country" in 1825, with eighty-three "volumes," [*sic*: probably unbound parts], whereas the actual title page states forty-one volumes. Precise dating

of components of the work is of interest mainly in establishing potential availability of its information to readers: e.g., was the "Cipher" article in the library at West Point when cadet Edward Porter Alexander, 1835-1910, USMA Class of 1857 (who introduced the Vigenére into Confederate service) might have had an opportunity to see it? (In characterizing the Blair extract as "the first U.S. government manual on cryptography," I am reserving "first American" for the Confederate confidential pamphlet (Richmond, 1862), compiled by Alexander and his brother, until that distinction is successfully challenged. No copies of that work, the Confederate equivalent of Myer's *Manual*, are known to exist today. See Friedman's Lecture IV in *The Friedman Legacy*, Center for Cryptologic History, 1992, or SRH-004, Record Group 475, National Archives.)

4. W. K. Wimsatt, Jr., "What Poe Knew About Cryptography." *Publications of the Modern Language Association of America*, LVIII, 3 (September, 1943), 754-779, supersedes William F. Friedman, "Edgar Allan Poe, Cryptographer" (*American Literature*, VIII (November, 1936), 266-280. See also Kahn, *Codebreakers*, 783-793. Both Wimsatt and Kahn had consulted with Friedman to gain his insights. Overseas, British Admiral Sir Francis Beaufort ("who was himself a great influence in developing and modernizing cryptography for Naval Intelligence") is quoted as having said that the Rees-Blair article "attracted much attention in London intelligence circles and then came the influence of Edgar Allen [*sic*] Poe" (Richard Deacon, *A History of the British Secret Service*. New York: Taplinger Publishing Co., 1969, 143).

5. For example, S.H. Lockett, "The Defense of Vicksburg" in Robert Underwood Johnson and Clarence Clough Buel, eds., *Battles and Leaders of the Civil War* (New York: The Century Co., 1884, 4 vols.), Vol. 3, "Retreat from Gettysburg," 482-492, at 492: "Our signal-service men had long before worked out the Federal code on the principle of Poe's 'Gold Bug,' and translated the messages as soon as sent."

6. See William Plum, *The Military Telegraph During the Civil War in the United States* (Chicago: Jansen McClurg & Company Publishers, 1882, 2 vols.; reprinted, New York: Arno Press, 1974, with introduction by Paul J. Scheips), 44 ff., and David Homer Bates, *Lincoln in the Telegraph Office* (New York: The Century Co., 1907), 49 ff. The author is persuaded that the principal appeal of the cipher was that the telegrapher was able to deal with recognizable words, instead of "sound" copying of Morse, and, confronted with line interference, this was an all-important consideration. The system was one "of, by, and for" telegraphers, rather than having been devised by a cipher expert. Through trial and error, it evolved during the war, and was approaching two-part code status at war's end, by a variety of variant routes. Post-1865 replacements evidently continued to serve government (or, at least, War Department) cryptography into the seventies, and probably until the appearance of the 1885-86 telegraphic code compiled under the direction of Lt. Col. J.F. Gregory, aide to commanding general Phil Sheridan, based upon the 1870 commercial code of Robert Slater (London).

7. See Max L. Marshall, ed., *The Story of the U.S. Army Signal Corps* (New York: Franklin Watts, Inc., 1965). Myer is considered to have based his signal code on the two-element Bain telegraphic code, with which he was familiar as a young telegrapher, rather than the four-element American Morse system. On the other hand, the Myer code (a simple substitution cipher, cryptographically) was similar to one depicted by Blair as an example of a two-or three-element system. (See 26 of *Cipher*.)

8. Adolphus Washington Greely (1844-1935), Chief Signal Officer from 1887 to 1906, was a soldier-scientist of distinguished career, commencing with Civil War service as a volunteer and culminating with his receipt of a rare "lifetime" Medal of Honor as major general. Meteorologist-climatologist, polar explorer, founder of the National Geographic Society, Greely was also, as chief signal officer, compiler of the 1906 War Department Telegraphic Code, among other accomplishments.

9. The United States Weather Bureau stemmed from the systematized recording,

reporting (via telegraph), and study of weather conditions commenced by the Signal Corps after the Civil War. Gen. Myer's intense interest in the prospect of predicting the weather led to the somewhat derisive nickname, "Old Probabilities." The weather service function was transferred from the Signal Corps to the Department of Agriculture in 1891.

10. Information provided to the author by Ms. Rebecca Raines, signal corps historian, Center of Military History, May 1992.

11. The possibility that source and/or authorship were unknown to the "extractor" cannot be ruled out: see Footnote 1 above. Roes evidently is "rediscovered" at intervals: In 1970, two extracts comparable to Cipher appeared, i.e., Rees's Clocks, Watches and Chronometers, 1819-20: a Selection from *The Cyclopaedia: of Universal Dictionary of Arts, Sciences and Literature* (Rutland, VT: C. E. Tuttle Co., 295 pp.) and Rees's *Naval Architecture (1819-20)* (Annapolis: United States Naval Institute, 183 pp.). As a basic repository of information, the Blair article offers a satisfactory answer to the question posed back in the sixties by the author to the late Lambros Callimahos: "How do you suppose the Confederates hit upon the Vigenére for their main cipher – how would an American in 1861, striking out on his own, have known about that system?" His response was, "Oh, probably from some encyclopedia, ..." which may well have been the case.

Chapter 15

Nicholas Trist Code

Fascinating dispatches, frequently encoded, from American ministers in Mexico flooded the State Department in Washington in the decades after May 1825 when Joel Poinsett arrived in Mexico and opened diplomatic relations with the weak and struggling Mexican nation. European intrigue, domestic plotting, economic depression, tensions between church and state, and suspicions about American Manifest Destiny kept the Mexican government in constant turmoil and tension. Poinsett noted how Mexican society had been dwarfed by Spanish colonization policies, and he accused the Spanish government of stunting the growth of Mexican society, leaving it closer to the Age of Charles V, thus 300 years retarded.

Poinsett's instructions from Secretary of State Henry Clay specified that he obtain a trade treaty and also a boundary agreement from the Mexican government. Soon after his arrival, he began negotiations on these objectives; however, he faced intense hostility from certain Mexican political leaders who sought a closer relationship with Great Britain and the other former Spanish colonies in the Americas. Clever, ambitious, politically sensitive, and devious, Poinsett organized political groups in Mexico to support the United States foreign policy objectives. A prolific writer, he reported on his imaginative programs and Mexican troubles, with many paragraphs in the Monroe code, first used in 1803, to Clay. Despite his energetic and highly secret maneuvers, Poinsett failed to secure the trade and boundary treaties before the Mexican government demanded his recall in 1830. His successor, Anthony Butler, pursued similar objectives and also wrote lengthy encoded dispatches; however, he did obtain a trade treaty with Mexico. After five years of anxious negotiations, Butler suffered the same fate as Poinsett and was recalled at the request of the Mexican government.

Future United States ministers in Mexico, such as Powhatan Ellis, continued to write their highly confidential dispatches to Washington in code. And in the decade after 1835, severe stresses between the two nations escalated, especially over the status of Texas and also the desire of United States (especially exhibited by President James Polk) for Mexican lands along the California coast. In addition, financial claims by American citizens for ships and goods seized by Mexican bureaucrats added to the threat of violence. War erupted in May 1846, and American armies under General Zachary Taylor invaded Mexican lands during the remainder of that year. The next year General Winfield Scott carried out a brilliant amphibious assault on Vera Cruz and advanced on Mexico City.

Nicholas Trist, chief clerk of the State Department, went to Mexico as Polk's secret agent in an effort to end the war. Born in Virginia in 1800, schooled for a time at West Point, and tutored in the law by Thomas Jefferson (and later, executor of Jefferson's will), he would see diplomatic service as U. S. consul in Havana, Cuba, between 1833 and 1841. Traveling to his new assignment and writing to James Buchanan, secretary of state, from New Orleans in late April 1847, Trist did not use code. His five subsequent letters to Buchanan from Vera Cruz and Jalapa during May were also written without code or cipher. His sensitive mission to secure peace with Mexico went badly, partially the result of a minor misunderstanding with General Winfield Scott, whose army he accompanied from Vera Cruz, and also because Trist's cover ended when newspapers published accounts of his secret mission, to the dismay of Polk and Buchanan.

On 3 June 1847, Trist wrote to Buchanan from Pueblo and explained his design for a code. "I have been occupying part of my time here in making a cypher, which I shall probably have frequent occasion for. A duplicate & key can be made at the Department by sending to my daughter for a copy of the smallest of the books (there are several at my house) that she packed up for me: the work of our old instructor, who was sent to Spain as Consul. Let the address of the prefatory address to the British Nation (excluding this title) be numbered from one onward until every letter of the alphabet is reached, except z (which I represent by zero). Each of the letters, with a few exceptions has three numbers corresponding to it."[1] Trist's code design was identical to that established by Charles William Frederic Dumas almost seventy-five years earlier.

above, second part, he used three numbers. "Three numbers between brackets will indicate the page, the line, and the letter. . . . If there be more than three numbers within the same brackets, all after the third will indicate letters in the same line. For example: (33,4,5) will indicate the letter g." This additional Trist system is similar to the earlier *Entick's Dictionary* code designs.

Which book Trist used as a basis for his messages remained a fascinating mystery for historians until several years ago. Intensive research by Stephen M. Matyas on books published by the approximately fifty American diplomats in Spain and Spanish dependencies before 1847 finally led him to suspect that Joseph Borras, the consul at Barcelona in 1836, might be the author described by Trist. He investigated Borras's small book, *Verdaderos principios de la lengua castellana* (True Principles of the Spanish Language) and found the prefatory address in the book began with "The study of foreign languages.. . ." This was indeed the source book used by Trist. One of the few extant copies is located in the Newberry Library.[3]

Trist's passage for the code began, "the study of foreign languages after ..." and had over 500 elements:

1	t	13	r	25	e	39	i	73	h	167	w
2	h	14	e	26	s	40	s	75	m	183	p
3	e	15	i	27	a	41	i	99	b	196	m
4	s	16	g	28	f	44	o	100	l	200	v
5	t	17	n	29	t	45	n	102	w	215	b
6	u	18	l	30	e	50	r	105	c	409	k
7	d	19	a	31	r	52	w	110	f	448	v[2]
8	y	20	n	33	h	58	d	114	p		
9	o	21	g	34	i	61	b	115	y		
10	f	22	u	35	m	64	d	121	m		
11	o	23	a	36	c	65	l	133	p		
12	o	24	g	38	u	66	y	147	t		

In his July 23 letter to Buchanan, Trist incorporated another design for writing "passages of special consequences" by going to a book code: using the same source book as

Trist's dispatches to Buchanan in code during these months reveal a regular procession of intense negotiations and maneuvers to secure an end to the armed conflict. President Polk changed his mind about securing a moderate treaty and instead became committed to demanding even more Mexican territory. Buchanan recalled Trist; however, Trist ignored the command and instead dedicated himself to securing a compromise treaty that would be moderate and save Mexico from collapse and anarchy.

Trist's skill in using the code did not match the interesting rhetoric of his letters. Based on his instructions, 78 words or 510 letters from the Borras's book were required in order to include all letters in the alphabet, except Z, which is not in the prefatory section. The distribution of numbers for the frequency of the letters of the alphabet is as follows:

A	37	N	43
B	9	O	44
C	19	P	8
D	13	Q	1
E	67	R	29
F	13	S	38
G	13	T	48
H	25	U	1
I	40	V	2
J	1	W	7
K	1	X	1
L	14	Y	11
M	10	Z	0

In actuality, he used only about sixty of the code numbers, mainly numbers "1" to "41" together with nineteen others. This limited use meant a dispatch intercepted by the enemy could, through item analysis, be read rather quickly. Had the full range of 510 code numbers been used, the secret communications would have been far more secure.

Trist's last encoded dispatch, written on 17 October 1847, marked the end of secret writing by American diplomats in Mexico before 1876. The treaty terms he negotiated in careful compliance with his original instructions became the Treaty of Guadalupe Hidalgo in February; the treaty was approved by the Senate in the spring of 1848 and ratified by the Mexican Congress two and one-half months later. For ignoring Buchanan's order to return to the United States months earlier, Trist came back to Washington in disgrace and also lost his State Department job; moreover, he was not paid for his time in Mexico. Eventually he took a clerking job with the Wilmington and Baltimore Railroad, and in 1870 became postmaster in Alexandria, Virginia. The next year he finally received almost $15,000 in salary and expenses from the government for his earlier special agent assignment in Mexico.

Notes

1. Nicholas Trist to James Buchanan, Pueblo, 3 June 1847, in *Dispatches From the United States Minister to Mexico, 1823-1906*, Record Group 59, Microcopy 97, Roll 15.

2. Fragments of the cipher have been determined from the following Trist dispatches to Buchanan: 13 June, 13, 23,31 July, 14,24 August, 4 September, and 17 October 1847: ibid., Roll 15. Also Trist to Hetty Parker, Mexico, 4 December 1847, Trist Papers, Library of Congress.

3. Albert C. Leighton and Stephen M. Matyas, "The Search for the Key Book to Nicholas Trist's Book Ciphers," *Cryptologia* 7 (October 1983): 297-314.

Chapter 16

Internal Struggle: The Civil War

David W. Gaddy

The greatest threat to the survival of the young republic came not from an external foe but from internal division. The secession of South Carolina in December 1860, followed by other Southern states and the formation of a rival Confederate States of America early the following year, left the Northern states in possession of the capital in Washington but bereft of the talent and territory that "went south." The four-year struggle that ensued was extraordinary in several respects. At the outset there were few, if any, secrets. Southerners had been at the seat of power for decades. For example, former Mississippi senator Jefferson Davis, chairman of the Senate Military Affairs Committee prior to his resignation, had been an outstanding secretary of war under President Franklin Pierce. West Point trained, a combat hero of the War with Mexico, he took his experience into the presidency of the rival Confederacy. His counterpart, Abraham Lincoln, had no comparable qualifications but possessed talents and ability that would well serve the Union cause, as well as a competent cabinet. One was locked into patterns of the past; the other was a "quick study."

The contending forces spoke the same language, shared the same social institutions, including an unmuzzled press and a tendency to express oneself freely on any subject. Military knowledge was shared in common — former classmates at the service academies and peacetime friends would meet in battle. They knew each other's strengths and weaknesses, and they eagerly devoured reports, in the press and through intelligence sources, of the names of opposing commanders. Each harbored sympathizers with the other side, the basis for espionage and a potential fifth column. Neither inherited any competence in information security nor an appreciation for operational security. Those things would be learned the hard way — the American way — accompanied by bloodshed.

From the standpoint of communication technology, the mid-nineteenth century had seen the introduction of electromagnetic telegraph — the inventions and variations of Morse, Bain, and others since the 1840s — and an initially abortive attempt at a transatlantic cable. British forces in the Crimea had used the telegraph for strategic lines, but mobile, operational use was a Civil War innovation. Of comparable importance to military communication technology and cryptology, a point-to-point visual system (a flag by day, torches by night) had been adopted by the War Department just prior to the war. Whereas the wire telegraph required physical contact for interception, the visual system, devised by an assistant army surgeon, Albert James Myer of New York, was susceptible to anyone else who could see the signals.

The Myer system, later known as wig-wag, featured a single flag, waved to the left or right somewhat like the binary dot-and-dash of today's International Morse Code. The flag used was selected to contrast with the signalman's background, as viewed from the distant point, usually a white flag with a red square or a red flag with a white square. At night the flag was replaced with a torch (a second torch was placed at the feet of the operator to serve as a reference point). These primitive implements of the first practical tactical military system of telecommunication are recalled even today in the insignia of the Army Signal Corps. They remained in supply until at least the First World War as an alternate means of signaling Morse. They were light, easily improvised, and cheap. With range extended through relays and the use of telescopes, the Myer system lent itself to a hierarchical overlay of the command structure in some instances (enabling an early form of "traffic analysis" and anticipating the advent of wireless telegraphy later in the century). As a companion to Morse, they made an excellent combination for the time. Wiretapping, signal interception and exploitation, authentication or identification systems (countersigns), the "war of wits" between "codemaking and codebreaking" for Americans truly stemmed from the American Civil War; and Myer's system, as well as the organizational concept of a corps of trained communicators, made an impact on other armies of the world.

One of the many ironies of the American Civil War was that a colleague of Myer, detailed to assist him in perfecting his system to the satisfaction of the War Department just before the war, served the South. Edward Porter Alexander of Georgia was a West Point graduate military engineer. With the secession of his state, Alexander opted for the Southern cause and was charged by President Jefferson Davis with organizing a signal corps to serve Confederate forces facing Washington. With the advance of the Federal army toward Manassas (Bull Run), Virginia, in the spring of 1861, Alexander found himself on an elevation that afforded an excellent view of the developing battle. A chance glance revealed a critical enemy flanking movement that endangered his side. Grabbing a flag, he frantically waved to attract the attention of one of his trainees: "LOOK TO YOUR LEFT - YOU ARE TURNED." The tactical warning, as well as demonstration of tactical communication, made quite an impression on generals of both sides. Myer subsequently used Alexander's exploit to petition Congress to organize a signal corps for the North, even as his erstwhile student and colleague established the system in the South.

Edward Porter Alexander

Sharing the same operational concept for signaling, Myer and Alexander prudently changed the basic code, or "alphabet," as it was generally called at the time, from the one the two men had originally used (based on the binary Bain code, with which Myer was personally familiar, as opposed to the four-element code of American Morse). They settled down in the fall of 1861, until observant members of both organizations came to realize, through close observation, that the flag "code" was actually a simple substitution cipher, and that, by applying the rules of Poe's "The Gold Bug," it could be readily broken. Thus began American signals intelligence and the "war of wits," as each struggled to read the other and protect his own signals. Wiretapping and manipulation added another dimension to the cryptologic war.

American cryptography of the period was little advanced from that of the Revolutionary era. The Confederate leadership initially fell back on the old "dictionary" cipher, in which the correspondents agree on a book held in common (generally a dictionary, both for vocabulary and convenience of arrangement) and designate plain text by substituting the page and position from the book. Simple ciphers abounded, some with mysterious-looking symbols instead of letters, presumed to offer greater security. In the North, as telegraphers (frequently little more than teenage boys) were pressed into service and formed into the U.S. Military Telegraph (USMT), a rival of Myer's signal corps, a word, or route, transposition system was adopted and became widespread. It gave the telegraphers recognizable words, an asset in this early stage of copying Morse "by ear," that helped to reduce garbles. Covernames or codewords replaced sensitive plain text before it was transposed, and nulls disrupted the sense of the underlying message. Only USMT telegraphers were permitted to hold the system, thereby becoming cipher clerks as well as communicators for their principals, and the entire organization was rigidly controlled personally by the secretary of war. In the War Department telegraph office near the secretary, President Lincoln was a frequent figure from the the nearby White House, anxiously hovering over the young operators as they went about their work.

In the South, although a Confederate States Military Telegraph was organized (in European fashion, under the Postmaster General), it was limited to supplementing the commercial telegraph lines. ("System" would not convey the proper idea, for the Southern lines were in reality a number of independent operations, some recently cut off from their northern ties by the division of the nation and reorganized as Southern companies.) Throughout the war, the Confederate government paid for the transmission of its official telegrams over commercial lines. Initially the Southern operator found peculiar digital texts coming his way (the dictionary system), then scrambled, meaningless letters, begging to be garbled. The polyalphabetical cipher used for official cryptograms offered none of the easily recognizable words that provided a crutch for his Northern brother.

A number of events and circumstances had led to this primitive attitude toward communication security. In the era before the war, the public had been fascinated with the deciphering of Egyptian hieroglyphics by Champollion (1790-1832), resulting from the discovery of the Rosetta Stone by Napoleon's troops in Egypt in 1799. The flourishing rediscovery of the mysteries of ancient Egypt was reflected in the growth of fraternal organizations, with secrecy, symbols, and ciphers part of their appeal. The popularity of Poe's writings has already been noted. The cost of telegraphy would spur interest in commercial codes that would reduce costs. And perhaps some then living had perused Blair's article in Rees's *Cyclopaedia*, but lacked the incentive to exploit it.

Against this background of innovation, embryonic technology, and innocence, the Civil War stands as something of a watershed, and the seeds and sprouts of cryptology are evident at every turn. In 1862, the South adopted the centuries-old Vigenére as its principal official cipher, then proceeded to violate its inherent strengths for the time by such practices as retaining plaintext word length, interspersing plain text and cipher, etc. A cipher that Alexander (who introduced the system through a pamphlet produced by his brother) had anticipated would be used with care, and primarily for short messages, was abused in the worst way, and Southern telegraphy compounded the problem of communication by garbling the cryptograms. Confronted with the knowledge that the enemy was reading his signals, the two sides initially reacted similarly, by changing the basic code , complicating life for themselves by losing the letter frequency association with the simpler signals. By 1863 the two sides went in different directions. The South went "offline," enciphering important messages with the Vigenére, then transmitting with a flag code that might or might not be "read" by the enemy. The North, on the other hand, adopted a handy "online" means of changing the basic flag code by prearrangement or at will, even within the act of transmission. This was done with a disk, in which the alphabet on the inner disk revolved against an outer ring of flag combination, enabling an instant change of code. With each year of the war, the two sides became more sophisticated, and yet, within weeks of Appomattox, each was still able to exploit the communications of the other – at least, at times – while secure in the belief that the other side could not possibly read friendly signals.

Although William Blair's turn-of-the-century essay, "Cipher," would have familiarized the reader with word transposition, or route cipher, as it was known in the 1860s, the man credited with introducing the system into American usage was Anson Stager – hardly a "household" name even among professional cryptologists or Civil War scholars, but deserving of recognition. The small band of USMT telegrapher-cipher operators who had never heard of the Duke of Argll knew their cipher only as Stager's. It was a system of, by, and for telegraphers. And, although scornfully disparaged by "the father of American cryptology," William F. Friedman, it served its purpose – which is about all one can ask of a cryptosystem.

Stager, a New Yorker (like Myer), was born in 1833. He began his working life as a printer's devil in an office under Henry O'Reilly (who became a leader in telegraph construction and management), then bookkeeper for a small newspaper before becoming a telegraph operator in Philadelphia and later Lancaster, Pennsylvania. Rapid promotion followed: after a brief time as telegraph office manager in Pittsburgh, he became, in his early thirties, general superintendent of the Western Union Telegraph Company, with headquarters in Cleveland, Ohio. Stager's early employment made him sympathetic with newsmen and their relations with the telegraph companies. Also, he convinced railroad executives that their companies could profit handsomely by permitting his company to share use of the railroad telegraph lines.

Soon after the outbreak of the Civil War, Stager took over responsibility for all of the telegraph lines in the Ohio military district, which placed him in association with a recent railroad executive who was also a West Point graduate and general of volunteers, George B. McClellan, who was to become prominent in the second year of the war for the Union cause. Stager made up a simple version of word transposition for the governor of Ohio to use in communication with the chief executives in Indiana and Illinois. At General McClellan's Cincinnati home, Stager provided him with a similar system, to be used between the general and detective Allan C. Pinkerton, whom McClellan employed as his intelligence and secret service chief. Stager accompanied McClellan's forces and established the first system of field telegraphs used in the war: "The wire followed the army headquarters wherever that went, and the enemy were confounded by the constant and instant communi-cations kept up between the Union army in the field and the Union government at home."[1] When the president took control of all of the telegraph lines in the northern states, telegraphers from the commercial lines and railroads were brought into government service, Stager among them. Loosely organized at first, the U. S. Military Telegraph was placed under Stager. Initially its members were contract employees of the War Department, but, as civilians in a combat area, their status was brought into question. They were issued uniforms, but with no insignia of rank – were they to be saluted or given orders? Finally, they were brought into military service under the quartermaster general (albeit under the direct control of the secretary of war), and Stager was commissioned a colonel. His route ciphers became the only accepted system for the USMT, and, since its operators were assigned to virtually all general officers, were the "mainline" or general Union cipher, the

Signal Corps' transmission security encryption notwithstanding.

In his "Lectures,"[2] William F. Friedman said, "I know no simpler or more succinct description of the route cipher than that given by one of the USMT operators, J.E. O'Brien, in an article in *Century Magazine,* XXXVIII, September 1889, entitled "Telegraphing in Battle":

> **The principle of the cipher consisted in writing a message with an equal number of words in each line, then copying the words up and down the columns by various routes, throwing in an extra word at the end of each column, and substituting other words for important names and verbs.**

"A more detailed description in modern technical terms," Friedman continued, "would be as follows: a system in which in encipherment the words of the plaintext message are inscribed within a matrix of a specified number of rows and columns, inscribing the words within the matrix from left to right, in successive lines and rows downward as in ordinary writing, and taking the words out of the matrix, that is, transcribing them, according to a pre-arranged route to form the cipher message." Friedman also noted that, while the basic principle, that of transposition, makes the "Stager" system a cipher, its incorporation of codewords (or "arbitraries," as Stager called them) makes it "technically a code system as defined in our modern terminology" (or simply "cryptosystem," to avoid being more definitive). Among its features, the system also employed what Stager termed "blind" or "check" words – nulls, we would say. These were generally placed at the end (top or bottom, depending on direction) of a column, signifying "turn here." Blind words also distorted the true dimensions of the matrix. "Commencement words," we would call them indicators, or key words, placed in the first group of the cipher text indicated the dimensions of the matrix used and/or the route pattern to be followed.

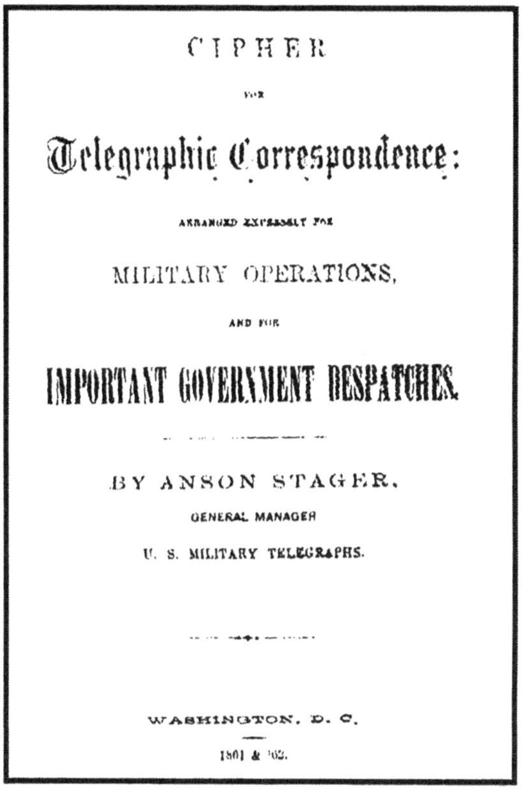
CIPHER

FOR

Telegraphic Correspondence:

ARRANGED EXPRESSLY FOR

MILITARY OPERATIONS,

AND FOR

IMPORTANT GOVERNMENT DESPATCHES.

BY ANSON STAGER,

GENERAL MANAGER

U. S. MILITARY TELEGRAPHS.

WASHINGTON, D. C.

1861 & '62.

Title page for Stager's cipher manual

Friedman was harsh in his assessment of Stager's creation. It was "utterly devoid... of the degree of sophistication one would be warranted in expecting in the secret communications of a great modern army in the decade 1860-1870, three hundred years after the birth of modern cryptography in the papal states of Italy." He found it improbable that the Confederates could not readily exploit it, and preferred to credit them with superior security that hid their success. He did concede to Stager's system some surprisingly modern features, features that recall the background of its prominence in a printing shop as well as telegraphy.

As skill developed, the practitioners freely indulged in phonetic or intentional misspelling of words, somewhat akin to today's "cablese" or ham slang, but intended as much to confuse the outsider as to communicate with the initiated. They also introduced more and more code word equivalents for personalities, places, dates and time, and the vocabulary of battle, even to the extent of brief phrases. Friedman realized that code books were printed with the plaintext equivalent blank, to facilitate reallocation without reprinting. (He seems not to have appreciated the fact that, in addition to the eleven or twelve "mainline" codes known to him, there were numerous lower level or departmental codes, not used with Washington, but controlled through the USMT.)

His study revealed that words and contemporary names expected to appear in a military context were intentionally used as code words or indicators to confuse a would-be analyst (whose approach to solution would be closely akin to anagramming). He noted a "two-letter differential " in the selection of code words, "a feature found only (otherwise) in codebooks of a much later date." "This principle," he stated, "is employed by knowledgeable code compilers to this very day, because it enables the recipient of a message not only to detect errors in transmission or reception, but to correct them." He noted that indicators and code words were prescribed with variants and that they were not in alphabetical order, and concluded that "these books partake somewhat of the nature of two-part or 'randomized' codes, or, in British terminology, 'hatted' codes." "The compilers of the (USMT) code books must have had a very clear idea of what I have just explained, but they made a compromise of a practical nature between a strictly one-part and a strictly two-part code, because they realized that a code of latter sort is twice as bulky as one of the former sort, besides being much more the laborious to compile and check the contents for accuracy."

Although Friedman noted that "it is to be remembered, of course, that messages were then transmitted by wire telegraphy, not by radio, so that enemy messages could be obtained only by "tapping" telegraph wires or capturing couriers or headquarters with their files intact," one wonders whether his harsh judgment of the Stager system was not based on later, radio era, considerations than those of the time. There were several instances in which Southern officers came into possession of USMT books (which were thereupon replaced), thus the type of cryptosystem was presumably known to the Confederacy. The problem was lack of volume: this was not the radio era. Interception was hit or miss, for the most part. Codes were localized. Perhaps the best example of the Confederate perplexity is afforded through the experience of E.P. Alexander, the father of the Confederate Army Signal Corps, who, in mid-war (1863) was handed a Union cryptogram taken from a captured courier and asked if he could read it.[3] One message, on the spot. Alexander knew he was confronted by a word transposition (recalling it for his family, he referred to it as a sort of "jumble," a charming and apt term) and by seizing on a local place name of two parts not afforded code equivalents he tried anagramming, but to no success. Here is his account:

"At this camp, I remember, one night just as I was going to sleep, particularly tired & sleepy, a courier from Gen. Bragg brought me a cipher dispatch captured from the enemy on its way up to Gen. Burnside at Knoxville; with the request that I would try & decipher it. It was a letter of 157 words all in ajumble beginning as follows:

> **To Jaque Knoxville, Enemy the increasing they go period this as fortified into some be it and Kingston direction you up cross numbers Wiley boy Burton & if will too in far**

strongly go ought surely free without your which it ought and between or are greatly for pontons front you we move as be stores you not to delay spare should least to probably us our preparing Stanton from you combinedly between to oppose fortune roanoke rapid we let possible speed if him that and your time a communication can me at this news in so complete with the crossing keep move hear once more no from us open and McDowell julia five thousand ferry (114) the you must driven at them prisoners artillery men pieces wounded to Godwin relay horses in Lambs (131) of and yours truly quick killed Loss the over minds ten snow two deserters Bennet Gordon answer also with across day (152)

"I had never seen a cipher of this character before, but it was very clear that it was simply a disarrangement of words, what may be called, for short, a jumble. Each correspondent, of course, had what was practically a list of the natural numbers, say from one up to 50, or whatever limit was used, taken in an agreed jumble, as for instance beginning 19, 3, 41, 22, &c. Then, the first word of the cipher would be the 19th of the genuine message, the 2nd cipher would be 3rd of message, the 3rd cipher would be 41st, &c.

"Now, it was quite clear that if the jumble covered only 75 or 50 words or less, it would have been used twice or more times in ciphering 157 words. If it were used twice or three times, I could, by comparison & trail, probably decipher the whole business. But if the jumble was not repeated, I could never decipher it without getting another message in the same jumble in order to compare the two.

"So my first task was to see if the jumble was repeated in the message. To do this, I first numbered all the words of the cipher, & then began to hunt for words which probably went together like 'according to' 'means of 'so that.' First I picked out as many of such likely pairs as I could find. Then, I would take one of these pairs, & note how many words separated them in the cipher. Then I would go over the whole cipher message, & see if, any where else, the same interval separated two words which would possibly make sense. If I could find such a pair, the interval between these two pairs might be the size of the jumble. Without going into more detail, it is enough to say that I worked on it the whole live long night, but every test showed that the jumble was not repeated. I found one pair of words which certainly belonged together, 'Lambs' & 'ferry' – for there was a 'Lamb's Ferry' on the Tennessee River. But it only made the demonstration absolute that the jumble was not repeated. I afterward found that the Federals made their jumbles by means of diagrams of rows & columns, writing up & down in different orders & then taking the words across; but the principles ofjumbling are the same, however it is mechanically done. And the safety of the message depends on the jumble not being repeated. They also used some blind words to further confuse the cipher. This made, indeed, a most excellent cipher, quick & easy, both to write & to decipher, which is a very great advantage. But there is one objection to it, in that it required a book, & that book might get into wrong hands."

The message Alexander tried unsuccessfully to unravel is as follows:

HEADQUARTERS DEPARTMENT OF THE CUMBERLAND

Chattanooga, October 16, 1863 - 7p.m.

Major-General Burnside,

Knoxville, Tenn.:

The enemy are preparing pontoons and increasing on our front. If they cross between us you will go up, and probably we too. You ought to move in the direction, at least as far as Kingston, which should be strongly fortified, and your spare stores go into it without delay. You ought to be free to oppose a crossing of the river, and with your cavalry to keep open complete and rapid communications between us, so that we can move combinedly on him. Let me hear from you, if possible, at once. No news from you in ten days. Our cavalry drove the rebel raid across the Tennessee at Lamb's Ferry, with loss to them of 2,000 killed, wounded, prisoners, and deserters; also five pieces of artillery.

Yours,

ROSECRANS[4]

Answer quick.

Here is how it came to be in the form that confronted Alexander (see figures 1, 2, and 3). Partly to conceal the true addressee, the message is addressed (in the usual style of the USMT) to the telegrapher-cipher operator serving him. In this case, it was Charles W. Jacques at Knoxville. "ENEMY" is a "commencement word" (system indicator) from Cipher No. 9 setting out a 10-line, 6-column transposition matrix with the plain text inscribed in the normal left-to-right manner, codewords ("arbitraries") substituted for sensitive names or terms as assigned. (Note that the system had not anticipated placenames such as Lamb's Ferry and Kingston, requiring that they be given "in the clear," and affording Alexander a modest crutch in the former case.) The pattern for extracting the transposition is to read down the third column (starting with "the increasing they. . . To" and adding a null or "blind word," in this case, "some," to frustrate cryptanalysis and to indicate "change pattern") up the fourth (with null, "boy" at the top), down the second, up the fifth, down the first, and up the fifth. This covers the first sixty words of the message. Parts 2 and 3 are treated as separate cryptograms. "STANTON" (there would, of course, have been no initial capitalization in telegraphic transmission) sets up a 6 X 6 matrix with the pattern commencing in a diagonal from the lower right-hand ("from") cell to the upper left-hand ("oppose"), where "fortune" is inserted as a pattern-changing null. The extraction continues down the first column with the codeword ROANOKE masking "cavalry" and SPEED as a null; up the sixth column, starting with "IF"; down the second; up the fifth; down the third; and up the forth. "MCDOWELL" is an alternative to STANTON, setting up the same matrix and pattern as used in Part 2, ending with JULIA indicating the time of origin of the message.

			BOY	GREATLY		
1	FOR	(BURTON) [BURNSIDE]	THE	(WILEY) [ENEMY]	ARE	PREPARING
2	PONTOONS	&	INCREASING	NUMBERS	ON	OUR
3	FRONT	IF	THEY	CROSS	BETWEEN	US
4	YOU	WILL	GO	UP	AND	PROBABLY
5	WE	TOO	•	YOU	OUGHT	TO
6	MOVE	IN	THIS	DIRECTION	AT	LEAST
7	AS	FAR	AS	KINGSTON	WHICH	SHOULD
8	BE	STRONGLY	FORTIFIED	AND	YOUR	SPARE
9	STORES	GO	INTO	IT	WITHOUT	DELAY
10	YOU	OUGHT	TO	BE	FREE	TO
	NOT	SURELY	SOME			

* Substituted codewords are in parenthesis. Words outside of the matrix are nulls.

Fig. 1. ENEMY (System Indicator)

Reviewing Alexander's explanation (and allowing for copying errors), it is evident that the numbers inserted in parentheses are his interpolation of word-count. He did not deduce that the message was in parts – actually, constituting three separate cryptograms of two different transposition patterns. He correctly paired "Lambs" and "Ferry" from his knowledge of a local placename, but his analysis was based on the assumption that a numerical relationship would yield a solution, which led to frustration and failure to solve the system.

	FORTUNE				THE	TIME
1	OPPOSE	A	CROSSING	AND	WITH	YOUR
2	(ROANOKE) [CAVALRY]	TO	KEEP	OPEN	COMPLETE	AND
3	RAPID	COMMUNICATION	BETWEEN	US	SO	THAT
4	WE	CAN	MOVE	COMBINEDLY	IN	HIM
5	LET	ME	HEAR	FROM	YOU	IF
6	POSSIBLE	AT	ONCE	NO	NEWS	FROM
	SPEED	THIS	MORE			

Fig. 2. STANTON (System Indicator)

	MUST				MINDS	HORSES
1	YOU	IN	TEN	DAYS	(OVER) [OUR]	(RELAY) [CAVALRY]
2	DRIVEN	THE	(SNOW) [REBELS]	ACROSS	THE	(GODWIN) [TENNESSEE]
3	AT	LAMBS	FERRY	WITH	LOSS	TO
4	THEM	OF	TWO	THOUSAND	KILLED	WOUNDED
5	PRISONERS	AND	DESERTERS	ALSO	FIVE	PIECES
6	ARTILLERY	YOURS	(BENET) [ROSECRANS]	ANSWER	QUICK	(JULIA) [7 P.M.]
	MEN	TRULY	GORDON			

Fig. 3. MCDOWELL (System Indicator)

Unless he had read about the Union system in a postwar account, he also inferred correctly that codewords ("blind words") had been employed.

Although he was an artillery commander by this time and perhaps a bit rusty, Alexander's case may be a fair indication of the state of Confederate ability in cryptanalysis — at least in the field and under unusual circumstances. With due respect to Friedman's disdain for the Stager-type of word transposition, the test of cryptosecurity is how well it holds up in its intended purpose, given the prevailing technology. On that score, it was a success.

The War Department cryptosystem of the Stager type was produced in nearly a dozen versions for top-level use. Regional commands (departments) had their own versions, generally simplified and localized, but conforming to the Washington pattern, and there may have been instances in which special versions were made up (as implied in the exchange between Grant and Halleck below). According to Plum, Stager's initial offer of a cipher was to enable confidential communication among governors in the mid-west. A slightly altered version, which Plum calls "the first one," was supplied to Pinkerton, the detective. War Department ciphers numbered 6 and 7 were used by the Union army in 1861, following the same basic scheme. If we term these (Nos. 6 and 7) Series I, then we have the following in consecutive wartime use:

Series II comprised Ciphers 12, 9, and 10

Series III comprised Ciphers 1 and 2

Series IV comprised Ciphers 3 and 4

(Note: Cipher 4 was the last wartime cipher. A post-conflict Cipher 5 was introduced on 5 June 1865.)[5]

Control over the USMT and its cryptosystems was absolute on the part of Secretary of War Stanton, making the whole system of secure communication a privacy system under his authority (indeed, the time spent by President Lincoln in the War Department "communication center," later recalled in a charming account by one of the young clerks,[6] may have been in part to ensure personal awareness of incoming and outgoing traffic). An example of Stanton's iron fist is afforded in an incident of 1864 in which even General Grant himself ran afoul of that control.

From his headquarters in Nashville, Tennessee, Grant notified General-in-Chief H.W. Halleck (the rough equivalent of today's Chief of Staff of the Army) in Washington by telegram on 20 January 1864:

I have ordered the cipher operator to give the Washington cipher to Colonel Comstock [of Grant's staff]. The necessity of this I felt whilst in East Tennessee, receiving dispatches I could not read until I returned. The operator received the following dispatch from Colonel Stager to Colonel [Samuel] Bruch [departmental head of the USMTI: 'Beckwith [Grant's telegrapher-code clerk] must not instruct any one in the cipher. An order will be issued and sent to you on this subject.

I protest against Colonel Stager's interference. I shall be as cautious as I possibly can, that improper persons do not get the key to official correspondence.[7]

Halleck responded to Grant by telegram the same afternoon: "The Secretary of War directs that you report by telegraph the facts and circumstances of the act of Lieutenant-Colonel Comstock, in requiring A.C. [*sic*: Samuel H.] Beckwith, telegraphic cipher clerk, to impart to him (Colonel Comstock) the secret cipher, entrusted to said Beckwith for use exclusively in your correspondence with the War Department and Headquarters of the Army."[8]

Grant replied the next day: "I ordered Beckwith to give Colonel Comstock the key to Washington cipher, in order that I might have always some one with me who had it. Whilst at Knoxville I experienced the disadvantage of not having given such an order before. I would recommend that a cipher be used not known to Colonel Stager or any operator."[9]

Colonel Stager's apologetic explanation to General Halleck is also dated 21 January:

The information furnished me led me to believe that the request of the staff officer for a copy of the cipher was without General Grant's authority, and as a new cipher had been arranged expressly for Mr. Beckwith's use at General Grant's headquarters, with the order of the Secretary of War recently issued that the operators for this duty should be held responsible for strict privacy in its use, I indited the message referred to, not thinking that it would come in conflict with General Grant's orders or wishes, the general having recently expressed his entire satisfaction with Mr. Beckwith's services.

I am exceedingly mortified at the result, as my only desire was to furnish the most reliable means of communication to General Grant with the War Department.

The new cipher was arranged with a view of being used by telegraph experts, and it is believed cannot be used with any success by others than telegraphers.

A great number of errors have been made by staff officers working ciphers, owing to their lack of experience in telegraphic characters, and it is believed that greater accuracy can be secured by placing ciphers in the hands of experts selected for this duty.

The new cipher differs in many respects from those formerly used, and the one arranged for General Grant should not be

known to any other party, hence my anxiety to keep it in Beckwith's hands.

I sincerely regret that General Grant is led to believe that it is willful interference on my part.[10]

Halleck informed Grant on 22 January 1864:

...It was known that the contents of telegrams communicated by means of existing ciphers have been made public without authority. As these ciphers have been communicated to a number of persons the Department was unable to discover the delinquent individual. To obviate this difficulty a new and very complicated cipher was prepared for communications between you and the War Department, which, by direction of the Secretary of War, was to be communicated to only two individuals, one at your headquarters and one in the War Department. It was to be confided to no one else, not even to me or any member of my staff.[11] **Mr. Beckwith, who was sent to your headquarters, was directed by the Secretary of War to communicate this cipher to no one. In obeying Colonel Comstock's orders he disobeyed the Secretary and has been dismissed. He should have gone to prison if Colonel Comstock had seen fit to put him there. Instead of forcing the cipher from him in violation of the orders of the War Department, Colonel Comstock should have reported the facts of the case here for the information of the Secretary of War, who takes the personal supervision and direction of the military telegraphs. On account of this cipher having been communicated to Colonel Comstock the Secretary has directed another to be prepared in its place, which is to be communicated to no one, no matter what his rank, without his special authority.**

The Secretary does not perceive the necessity of communicating a special cipher, intended only for telegrams to the War Department, to members of your staffany more than to my staffor to the staff officers of other generals commanding geographical departments. All your communications with others are conducted through the ordinary cipher. It was intended that Mr. Beckwith should accompany you wherever you required him, transportation being furnished for that purpose. If by any casualty be separated from you, communication could be kept up by the ordinary cipher till the vacancy could be supplied.

It is to be regretted that Colonel Comstock interfered with the orders of the War Department in this case. As stated in former instructions, if any telegraphic employee should not give satisfaction he should be reported, and, if there be a pressing necessity, he may be suspended. But as the corps of telegraphic operators receive their instructions directly from the Secretary of War, these instructions should not be interfered with except under very extraordinary circumstances, which should be immediately reported.

P.S. Colonel Stager is the confidential agent of the Secretary of War, and directs all telegraphic matters under his orders.[12]

Grant responded to Halleck on 4 February::

Your letter of the 22nd, inclosing copy of Colonel Stager's of the 21st to you, is received. I have also circular or order, dated January 1, 1864, postmarked Washington, January 23, and received on the 29th.

I will state that Beckwith is one of the best of men. He is competent and industrious. In the matter for which he has been dis-charged, he only obeyed my orders and could not have done otherwise than he did and remain. Beckwith has always been employed at headquarters as an operator, and I have never thought of taking him with me except when headquarters are moved. On the occasion of my going to Knoxville, I received Washington dispatches which I could not read until my return to this place. To remedy this for the future I directed Colonel Comstock to acquaint himself with the cipher.

Beckwith desired to telegraph Colonel Stager on the subject before complying with my direction. Not knowing of any order defining who and who alone could be entrusted with the Washington cipher, I then ordered Beckwith to give it to Colonel Comstock and to inform Colonel Stager of the fact that he had done so. I had no thought in this matter of violating any order or even wish of the Secretary of War. I could see no reason why I was not as capable of selecting the proper person to entrust with this secret as Colonel Stager: in fact, thought nothing further of the, than that Colonel Stager had his operators under such discipline that they were afraid to obey orders from any one but himself without knowing first his pleasure.

Beckwith has been dismissed for obeying my order. His position is important to him and a better man cannot be selected for it. I respectfully ask that Beckwith be restored.

When Colonel Stager's directions were received here the cipher had already been communicated. His order was signed by himself and not by the Secretary War. It is not necessary for me to state that I am a stickler for form, but will obey any order or wish of my superior, no matter how conveyed, if I know, or only think it came from him. In this instance I supposed Colonel Stager was acting for himself and without the knowledge of any one else."[13]

Having satisfied Washington, Grant received on 10 February a telegram from Halleck that stated, among other things unrelated, "Mr. Beckwith has been restored."[14]

The order delayed in reaching Grant was as follows:

WAR DEPARTMENT

Washington City,January 1st, 1864

ORDERED:

That the cipher issued by the Superintendent of Military Telegraphs be restricted only to the care of telegraph experts, selected for the duty by the Superintendent of Telegraphs, and approved and appointed by the Secretary of War for duty at the respective headquarters of the Military Departments, and to accompany the armies in the field. The ciphers furnished for this purpose are not to be imparted to any one, but will kept by the operator to whom they are entrusted, in strict confidence, and he will be held responsible for their privacy and proper use. They will neither be copied nor held by any other person, without special permission from the Secretary of War. Generals commanding will report to the War Department any default of duty by the cipher operator, but will not allow any staff officer to interfere with the operators in the discharge of their duties.By order of the Secretary of War

E.D. TOWNSEND,

A.A.G.

Official: T.S. BOWERS, A.A.G.[15]

A variety of simple or improvised forms of cryptography or signaling appeared during the course of the war. Union agent Elizabeth Van Lew in Richmond used a 10 x 10 dinomic substitution system (frequently sent on tiny slips of paper, obviously concealed in transmission).[16] Lincoln himself toyed with a reversal of plain text, combined with phonetic spelling.[17] And "clothes-line" signals conveyed simple messages, such as "the coast is clear" or "enemy here."

The U.S. Navy in the Civil War retained its traditional hoisted flag signals in prearranged code (a new book was issued in 1864), and, in what may well have been the earliest example of interservice or joint telecommunications between the army and navy, accepted Myer-trained army signalmen aboard ships to coordinate operations. This resulted in Myer adopting, and the navy accepting, a "General Service Code" for flag and torch that lasted until the 1880s, when the International Morse Code replaced it.[18]

The State Department, on the other hand, appears not to have used any form of encryption for its correspondence with emissaries abroad — meaning that no extra effort was required on the part of British or continental postal authorities to exploit such dispatches through their hands.[19]

To summarize the American experience (both North and South) with cryptography during the Civil War, the following table illustrates the variety:

Union Cryptography

I. Combined cipher/code cryptosystem: route or transposition (USMT); simple substitution encipherment in text

II. Cipher

- A. disk (Signal Corps, for visual signaling)
- B. dinomic substitution (Van Lew)

III. Miscellaneous (Lincoln's reversed phonetics; clothes-line, countersigns, signals)

Confederate Cryptography

I. Codes

- A. dictionary
- B. open code
- C. signs and signals

II. Ciphers

- A. substitution
  1. simple, monographic substitution
  2. simple, symbols
  3. simple, keyed
  4. polyalphabetic; Vigenère
- B. transposition: revolving grille

III. Concealment

- A. microdot
- B. ink
- C. compact notes

Notes

1. Anson Stager, *Cleveland, Past and Present; Its Representative Men: Comprising Biographical Sketches of Pioneer Settlers and Prominent Citizens with a History of the City*, (Cleveland: Maurice Joblin, 1869), 449.

2. National Security Agency, Center for Cryptologic History, *The Friedman Legacy: A Tribute to William and Elizebeth Friedman*. Fort George G. Meade, Maryland, 1992, 84ff.

3. Gary W. Gallagher. *Fighting for the Confederacy: The Personal Recollections of General Edward Porter Alexander* (Chapel Hill: The University of North Carolina Press, 1989), 302-03.

4. U.S. War Department, *The War of the Rebellion: A Compilation of the Official Records of the Union and Confederate Armies*, 128 vols. (Washington, D.C.: U.S. Government Printing Office, 1880-1901). Series I, Vol. XXX, Part IV, 428. Hereafter cited as OR (Official Records). Ibid., 459. Union authorities had been alerted to the possible interception. In a message of 18 October, "I have just learned that one of our couriers having a dispatch from Major General Rosecrans to General Burnside has disappeared. The dispatch has not been received here. I am sending out 25 men to search for him. Please notify General Burnside of the loss of the dispatch. I cannot learn yet whether the courier was captured or not..."

5. Plum, II, 346. Federal control over telegraphs in the former Confederate States ended 1 December 1865 and Superintendent Stager's final report dated 30 June 1866. One former USMT cipher operator remained on duty until 1869.

6. David Homer Bates, *Lincoln in the Telegraph Office* (New York: Century, 1907). Mr. Bates (1843-1926) lived into the twentieth century as an official with Western Union.

7. OR, Series I, Vol. XXXII, Part II, 150.

8. Ibid., 159.

9. Ibid., 161.

10. Ibid., 161.

11. Compare Halleck's 20 January telegram, which seems to imply that Army Headquarters also held this cipher.

12. Ibid., 172-73.

13. Ibid., 323-24.

14. Ibid., 361.

15. Plum, *The Military Telegraph During the Civil War*, 170-1. It may be mean spirited to conjecture that this circular was back-dated, but several considerations suggest the possibility for anyone familiar with bureaucratic procedures: Grant's 4 February 1864 letter to Halleck says that he received his copy for the first time on 29 January, bearing a Washington postmark of 23 January, which suggests leisurely dispatch of a presumably important policy. Colonel Stager's 21 January dispatch letter to Halleck refers to the order as "recently issued," which does not, of course, rule out the possible 1 January date. But the order is remarkable in that it covers all of the aspects of the Grant-Beckwith-Comstock incident. And it was not published in the Official Records.

16. A copy was later found in Miss Van Lew's watch case. See William Gilmore Beymer, *On Hazardous Service* (New York: Harper and Brothers, 1912) for an account of the Van Lew ring and her cipher.

17. Plum, 1, 35. The text was written in reverse as well.

18. The Myer code was briefly introduced into the schooling of cadets at West Point. . . until instructors began noticing the creative movement of a pen or pencil during examinations.

19. Weber, *United States Diplomatic Codes and Ciphers, 1775-1938*, 214.

Chapter 17

Seward's Other Folly: America's First Encrypted Cable

On the early morning of 26 November 1866, before the American minister to France, John Bigelow, was out of bed, a secret encrypted cable from Secretary of State William Seward began arriving in the Paris telegraph office. The dispatch's last installment was completed at 4:30 the following afternoon. "I immediately discerned," wrote Bigelow, "that it was written more for the edification of Congress than for mine, for Mr. Seward knew full well at the moment of writing it that the Emperor [of France] and his Cabinet were all more anxious than any citizen of the United States to hasten the recall of their troops from Mexico, and that they were doing everything that was possible to that end."[1] News and rumors about the lengthy encoded telegram spread rapidly through the French governmental departments and the diplomatic corps: legation representatives flooded Bigelow's office with inquiries. Bigelow maintained a determined silence. The first steamer from New York to arrive in France after the dispatch was written brought a reprint of the confidential cable in the pages of the *New York Herald.* A confident Bigelow smiled: the reprint "confirmed my first impression that it was written for Congress rather than for the Tuileries."[2]

This strange episode in American foreign relations commenced a fascinating chapter in American cryptologic history. Moreover, the event shaped American State Department codebooks for the next two generations and also precipitated a costly lawsuit against the United States government.

Several months earlier, in early August 1866, John Bigelow, the forty-nine-year-old American ambassador to France, wrote William Seward about the receipt of an inaugural dispatch from the Atlantic cable entrepreneur, Cyrus Field, who transmitted a special message from Newfoundland to Paris: "The Atlantic cable is successfully laid: may it

Cyrus Field

1436, one hundred nine, 109, arrow, twelve sixty.four, 1264, fourteen hundred one, 1401, fifteen forty-four, 1544, three sixty, 360, two hundred eight, 208, eleven hundred eight, 1108, five twenty, 520; five sixty.nine,569, ten sixty-eight, 1068, six fifty-three, 653,six sixty-eight, 668, fourteen forty, 1440, fourteen thirty-six, 1486, three sixty-six, 366, four seventy-nine, 479, seventy, 70, five sixty-nine, 569, eight forty.six, 846, four ninety-one, 491, cross, eleven seventy-three, 1173, thirteen eighty-five, 1385, seventy-eight, 78, ten forty-seven, 1047, nine hundred eight, 908, ten forty-seven, 1047, three sixty, 360, twelve fifty-nine, 1259, fifteen

Extract from Seward dispatch to Bigelow

prove a blessing to all mankind."[3] Bigelow also joined in singing the chorus of congratulations and praised what he termed the "umbilical cord with which the old world is reunited to its transatlantic offspring."

Bigelow's laudatory comments reflected somewhat his cosmopolitan personality, shaped not only by his education at Trinity College in Hartford, Connecticut, and legal studies at Union College in Schenectady, New York, but also his first public office as inspector of Sing Sing prison. Later, accepting an invitation from William Cullen Bryant, Bigelow became part owner and editor of the *New York Evening Post.* During his travels, while editor, in Jamaica, Haiti, and Europe, he initiated friendships with Professor Charles Sainte-Beuve, a French literary critic; with Richard Cobden and John Bright, members of Parliament who were free traders and critics of the Crimean War; and with the English novelist William Thackeray. His European travels brought a distinctly cultured attitude to his own writings and musings on diplomatic relations and added an acute awareness of European government practices regarding communications security.

Politically astute, Bigelow, who became consul-general in Paris in 1861 and minister in 1865, recognized the new challenges for communications security that accompanied the new Atlantic cable. He strongly advised Seward to develop a new cipher for the exclusive use of the State Department so that Seward could communicate secretly with his diplomatic officers; even better, he suggested a different cipher for each of the legations. He warned Seward, "It is not likely that it would suit the purposes of the Government to have its telegrams for this Legation read first by the French authorities, and yet you are well aware that nothing goes over a French telegraph wire, that is not transmitted to the Ministry of the Interior."[4]

More worrisome to Bigelow was his belief that the State Department code was no longer secret, for he believed copies of it were taken from the State Department archives by the "traitors to the Government under Mr. Buchanan's administration" and the principal European governments now had the key. In conclusion, Bigelow added, the department should take steps to "clothe its communications with that privacy without which, oftentimes, they would become valueless."[5]

Seward's naive reply to Bigelow's dispatch dismissed the conjecture that traitors took copies of the code by stating that the code sheets were always in the custody of the department's loyal chief clerk or clerk in charge of the French and other missions. Moreover, if a person were to make a copy, it would take a least two long working days if he had the necessary blank forms, and a least a week without the forms. Then Seward, continuing to write as a person who had never used the code, noted that a variation of a single figure or letter would spoil the whole code. And he added an astonishing statement: the Department code, in service for at least half a century, was believed to be the "most inscrutable ever invented."[6] Seward wrote that he, together with earlier secretaries of state, held this opinion, and therefore the Department rejected the offer of five or six new ciphers each year. Apparently, Secretary Seward's management skills did not include an understanding of

William H. Seward

communications security, especially in a European atmosphere.[7] Nor did he understand the administration of cable communications when codes or ciphers were involved. Bigelow thought Seward too talented and ambitious to be satisfied with being merely a political swashbuckler; rather the secretary tried to rank with the leaders of men. However, "his wings, like those of the ostrich, though they served him to run with greater speed, could not lift him entirely from the ground. . . . If he did not march as fast as some, he always kept ahead of his troops, but never so far that they could not hear his word of command."[8]

* * * * *

A festive celebration on 29 August 1866, organized by New York citizens at the gaily decorated Delmonico's Hotel, on Fifth Avenue and Fourteenth Street, in honor of President Andrew Johnson and his Reconstruction leadership, attracted a large gathering of Republican admirers, local politicos, business leaders, and at least fifty reporters. Earlier that afternoon, more than 500,000 persons, kept orderly by "blue-coated gentlemen with brass coated buttons and locust clubs" welcomed President Johnson, Secretary of State William Seward, Secretary of the Navy Gideon Welles, and Postmaster General Alexander Randall to the city. Exuberant crowds and multicolored flags bordered the gala parade route for the president, his official party and the 4,700 marching soldiers: battery guns and fireworks added sound and color to the celebration.[9] As the patriotic procession passed slowly along the crowded streets, and ended at the gaily decorated hotel, a huge banner was unfurled near the hotel:

THE CONSTITUTION

WASHINGTON ESTABLISHED IT

LINCOLN DEFENDED IT

JOHNSON PRESERVED IT

At the gala dinner in the banquet room, decorated with state flags, over 225 of New York's leading merchants, ministers, and politicians were seated. Johnson's after-dinner speech, interrupted numerous times by lengthy applause, emphasized the necessity for Congress to cooperate with the president in restoring the Southern states to their rightful place in the federal government. Troubled by tensions with the recalcitrant Congress, he explained, "I believe that the great mass of the people will take care of the Government, and when they understand it, will always do right. You have evinced a good will; the billows begin to heave, and I tell those persons that are croaking and seeking individual aggrandizement, or the perpetuity of a party, that they had better stand out of the way; the country is coming together again." [10] In another brief address, Seward repeated the restoration theme and concluded that the Civil War and its immediate aftermath had abolished slavery, repudiated the Southern debt, and abolished the principle of secession, now and forever. And since these three issues had been solved, no further reasonable objections could be made to the admission of Southern representatives.

At the conclusion of the evening's festivities, Mr. Wilson G. Hunt, one of the directors of the New York, Newfoundland, and London Telegraph Company, led his younger friend, Richard Lathers, a Republican businessman, to the head table to meet the president.[11] Hunt knew William Seward, the secretary of state, also seated at the head table, and asked him for an introduction to the president, who was then busy talking to another dinner guest.

Hunt took these five to ten minutes to ask Seward why the federal government did not use the new Atlantic cable, which had just been completed on July 28. It was a question that would eventually lead to a $32,000 claim against the United States State Department. Replying to Hunt, Seward said that the tariff was too costly, that "the Government of the United States was not rich enough to use the Telegraph."[12] And Seward's judgment, though exaggerated, was somewhat accurate because the provisional tariff rates, adopted 1 July 1865, were very expensive: cable charges between America and Great Britain were $100 or 20 pounds sterling for messages of twenty words or less, including address, date and signature: every additional word, not exceeding five letters, cost 20 shillings per word. Between America and Continental Europe charges were 21 pounds for twenty words. Code or cipher messages were charged double.[13] All messages, according to the tariff, had to be paid in gold before transmission.[14]

Seward explained to the sixty-five-year-old Hunt "the government was too poor to use the cable, because the charges for its use, according to a tariff which was reported, were too high, and practically oppressive and extortionate."[15] Seward alarmed Hunt when he declared, "under that tariff, the Atlantic cable would, as a medium of communication between governments in Europe and America, be a failure; that the United States government would not use it, and I had learned from foreign ministers residing in Washington that they could not use it."[16] Indeed, Seward explained, he had earlier prepared a message to send to one of the American ministers abroad, and referred it to the telegraph company for transmission; however, on learning the estimated charges, (Hunt believed Seward mentioned the cost at about $680) he cancelled the request and sent the dispatch by mail.[17]

In addition, Seward said, the immense Civil War debt facing the United States required economy and frugality. He was acutely aware that the federal government had spent over three billion dollars during the four years of conflict; moreover, the federal debt equalled almost one half of the gross national product. Government leaders faced the largest debt the United States had ever experienced: the interest alone surpassed the federal debt before 1861.[18] In fact, Seward's overseas budget had been recently reduced from $140,000 for the fiscal year ending June 1866 to $115,000 for 1867. The State Department, Seward added, would lose public confidence if it incurred the great expense of telegraphic communication under the existing tariff. Moreover, Seward recognized that a code or cipher must be employed for telegraphic communication in order to maintain confidentiality; and using the U.S. "cipher code" for a cable at the time "increased the number of words about five times, and the expense of transmission ten times."[19] Erroneously, Seward believed the State Department code then current was the only one used since the federal government had been organized.

At the Delmonico dinner, an anxious Hunt told Seward that the telegraph tariff had been adopted on the grounds of the cable's novelty, and also it resulted from managerial inexperience with setting rates. He urged Seward to convey the State Department's objections in a written communication to the company proprietors. Seward either promised or indicated he might do so, perhaps after further reflection and consultation with the president.[20]

Seward said he believed it was at this time that Hunt asked what rates the government paid the domestic telegraph company. Seward replied that the War Department "conducts that business exclusively" under regulations made by the War Department, that the "war telegraph was a war instrument, and as I understood it, we fixed our own prices and paid what we pleased."[21] However, Seward's understanding was mistaken, for the government paid regular rates on Western Union lines. According to Seward, Hunt asked whether Seward would use the Atlantic cable telegraph by way of trial in the same way as the domestic telegraph adaptation until some definite arrangement could be made satisfactory to all. Seward promised to use the cable when a proper occasion arose, and they both agreed that the government would do what was just, and he hoped the telegraph proprietors would be equally reasonable.

According to Seward's account, Hunt and he had the understanding that Seward could pay what he thought proper for the trial use of the cable, and, moreover, that Seward should either send the dispatch to Hunt's care or advise him that the cable had been given to the agent so that the trial message would not be sent under the regular tariff, but subject to the special trial arrangement. Lathers, who was standing near Seward and Hunt during this discussion, later recalled Seward's emphasis upon economy but when questioned further, Lathers had no re-collection of the trial message option. Nor did Hunt, in his later deposition, recall any special trial message arrangement.[22]

The after-dinner conversation between Hunt and Seward ended with Hunt's belief he would soon receive a written message from Seward with a request for lower rates. Seward, in turn, said he believed he could send a trial message as an experiment for lowering rates. The seeds of confusion, planted during this brief conversation, would grow when Seward failed to send the written communication to the company's proprietors.

Seward also had allies in his complaints about the exorbitant cable tariffs. An editorial in *The New York Times* praised the ingenuity that provided telegraphic communication between the two continents, an "achievement much more grand than the 'Hanging Gardens of Babylon' or any other one of the wonders of the Old World."[23] However, the *Times* added that this monopoly should not "bleed the people." This newspaper and other large east coast publications were eager to lower their costs for the cables sent to them by foreign correspondents. Prices, the editor wrote, must be lowered: $5 in gold per five-letter word is too expensive. And with pleasure, the *Times* reported six weeks later on a letter from Cyrus W. Field that on and after 1 November 1866, Atlantic cable rates would be reduced fifty percent.[24] Negotiations between the New York, Newfoundland, and London Telegraph Company and the Anglo-American Telegraph Company resulted in the lowered tariff: messages of twenty words for $50 to Great Britain, and $51.25 to Paris. Code and cipher messages would still be charged double.[25]

Wilson Hunt sent Seward a listing of the new prices. Ten days after the new tariff went into effect and to the delight of the cable company, Seward sent, in plain text, the very first State Department cable via the Western Union Telegraph Company. It was a brief dispatch to John Bigelow, the American minister to France, simply telling him that his successor, General John A. Dix, would embark on the Fulton on 24 November.[26] Although cable company rules required prepayment for all messages, the State Department did not pay the charges of $60.37 for twenty-three words until the following May.[27] Cable company directors now hoped the federal government would send frequent communications via the Atlantic cable.

* * * * *

On 15 November 1866, in New York City's Metropolitan Hotel banquet hall, 300 invited merchants, bankers, and other distinguished guests were seated at a grand festival dinner, organized by the Chamber of Commerce, honoring Cyrus W. Field for his outstanding work in the thirteen-year project for the laying of the Atlantic cable. It was, noted the Chamber president, a celebration of Field's great work in "uniting by telegraph the Old World with the New." And Field was a determined entrepreneur: he is said to have crossed the Atlantic sixty-four times on cable business – suffering from seasickness each time![28]

The banquet hall was transformed into a magnificent flower-laden temple with a miniature globe of the planet earth (with an iron band signifying the telegraph wrapped around it) suspended in mid-air. Above the globe were the sun, moon, and stars from which silken threads came to the globe, and from there to miniature telegraph poles of silver on each of the ten dinner tables. High in the miniature solar system hung signs "Greeting from all the stars" and "The moon her peaceful radiance lends"; and hanging from the globe was a large signboard, "General Telegraphic Office."

Also beneath the globe hung a "a crown of flowers, emblematic of the peaceful reign of power which, it is to be hoped, has at last begun."[29] And prominent was the American flag, together with those of Russia, Denmark, Italy, England, Austria and Prussia. Among the prestigious invited guests who honored Cyrus Field were General George Meade, John Jay (grandson of John Jay discussed in chapter 2), Archbishop John McClosky, Peter Cooper, Horace Greeley, Reverend Henry Ward Beecher, George Bancroft, and William Evarts. After the dinner, the dining room doors were opened and forty to fifty ladies joined the celebration, seating themselves in different sections of the hail, pleased to join the male audience before the formal toasts and speeches.

In his most absorbing narrative on the challenges facing the telegraph builders, Field recounted the tremendous difficulties over the previous thirteen years, especially for financing and constructing the complicated project that consisted of four telegraph lines: London to Valentia, Ireland; Valentia to Heart's Content, Newfoundland; Heart's Content to Port Hood, Nova Scotia; and Port Hood to New York City. He gave special gratitude to British financiers for their enormous support over the years even though over $1 million had been spent by New York investors for the western terminus of the cable before a penny had been spent in England for the project. He also emphasized his hope that it would take no longer than twenty minutes for messages to reach New York from London: indeed, he thought a message from Wall Street to the Royal Exchange in London could be answered and returned to New York in an hour, even by allowing ten minutes on each side for a boy to carry the dispatch from the telegraph office to the business office.

Sensitive to the press and private complaints about the costly, indeed oppressive, tariffs, Field explained that the investment totaled $12 million. The managers initially were worried that the cable might again break; in fact, Field reported, some prophets predicted it might last only one month. And now, the company had two cables instead of only one, and a third distinct line planned. Experience had shown that instead of five words a minute, operators could send fifteen. Thus, after only three months of operations the tariff was reduced by just one half, and he hoped it would soon be brought down to one quarter.

Despite the anxieties of some Americans[30] about the fact that both ends of the cable rested on English soil and under English jurisdiction, the indefatigable and ever-optimistic entrepreneur Field, reflecting his English heritage, closed his discourse with unqualified praise for England: "America with all her greatness has come out of the loins of England – and though there have been sometimes family quarrels – bitter as family quarrels are apt to be – still in our hearts there is yearning for the old home, the land of our fathers; and he is an enemy of his country and of the human race, who would stir up strife between two nations that are one in race, in language and religion. I close with this sentiment: England and America – clasping hands across the sea, may this firm grasp be a pledge of friendship to all generations."[31]

Following the lecture, letters from President Andrew Johnson, several cabinet members, General Ulysses S. Grant, Senator Charles Sumner and many other dignitaries who were unable to attend the celebration were read to the guests. William Seward telegraphed this rather awkward greeting to Field: "The first, most constant and the most energetic friend in the United States of the latest accomplished great enterprise in the advance of universal civilization."[32]

Wilson Hunt's earlier request to Seward for greater government use of the cable would be answered a week after the New York banquet in honor of Field. Threatening events in Mexico, where French troops supported a European emperor, forced Seward to consider sending a secret encrypted warning to the French emperor, Napoleon III. The continuing revolution and warfare in Mexico had troubled the secretary all during the American Civil War. He feared this new expansion of a French empire in America. And with the war's conclusion, the situation along America's southern border now became a major foreign policy problem confronting Seward.[33]

Seward believed it was necessary to send a dispatch to his minister in France, John Bigelow, encoded because his highly confidential message would pass through American and foreign telegrapher hands. However, encoded American diplomatic dispatches had become a distinct rarity in the years after 1848, the end of the War with Mexico. Indeed, the last encoded communication involving an American minister in a major European nation before 1866 came from Romulus M. Saunders, stationed in Madrid, Spain. Writing to Secretary of State James Buchanan in November 1848, Saunders, who had earlier employed English and French couriers to transmit his dispatches, now prepared part of his diplomatic message in code, and explained he decided to risk sending the message by mail because the expense of a courier was not justified in this instance.

Buchanan, impatient in his last months as secretary to purchase Cuba from Spain for $120 million, anxiously awaited word from Saunders regarding this offer. Using the old Monroe code sheets, first used in 1803, Saunders masked this sentence in his dispatch: "N.B. I have had no encouragement to renew the subject in regard to Cuba: so far as I have been able to collect the opinion of the publick it is against a cession; and I do not think the present ministry could or would venture on such a step both Pidal and monarch against it and Narvez says nothing."[34]

The decline of American encrypted diplomatic communications in the 1840s mirrored a new liberal tradition sweeping Great Britain during this time. In support of oppressed Polish leaders and others persecuted by Russia or Austria, British laws were amended, and asylum awarded to all foreigners. In 1844, Thomas Duncombe told the House of Commons that mail addressed to the Italian revolutionary Guiseppe Mazzini, then in London exile, had been opened at the Post Office. This dishonorable action, he recounted, was taken at the request of the Austrian ambassador in London, who feared Mazzini was planting seeds of revolution in a region of Italy under the control of Austria.

The tampering had come to light when Mazzini asked a loyal correspondent to place poppy seeds in envelopes: when Mazzini opened the envelope, the seeds were missing. A Committee of Inquiry looked into British mail opening practices and recommended "to leave it to mystery whether or no this power is ever exercised" and therefore "deter the evil-minded from applying the Post to improper use."[35] Thus, the law remained but the practice changed because in that same year, the secret foreign letter monitoring branch of the Post Office was abolished along with the deciphering office. However, the branch inspecting domestic mail appears to have continued.

During the American Civil War, French armed forces, under orders of Napoleon III, captured Mexico City and in 1864 arranged for Archduke Ferdinand Maximilian of Austria, then thirty-two years old, to take over the Mexican throne. A shrewd Secretary of State William Seward, anxious about potential French support for the Southern armies if he complained too vigorously about French intervention in Mexico, patiently waited until Southern military forces no longer threatened the Union.

In the months immediately after the South's surrender at Appomattox, the apprehensive Seward pressured Napoleon III to withdraw his military forces in Mexico, then numbering 28,000 men. According to Seward, this withdrawal would enable the Mexican people to choose between Maximilian as emperor and Juarez as president.[36] In January 1866, the French emperor ordered his military staff in Mexico, headed by Marshal Francois Achille Bazaine, to prepare for evacuation from Mexico. By April, the emperor agreed that 28,000 French troops would leave in three stages: November 1866, and March and November 1867.[37] In late May, Bigelow was told the French troops would be withdrawn, probably sooner than the scheduled time.[38] In June, Maximilian received word from Napoleon III that the French army was being sent home. In late August, press accounts stated that Napoleon had been visited by the Empress Carlotta, Maximilian's wife, recently arrived from Mexico. She requested an extension of the time for the departure of the French troops from Mexico, and Napoleon granted her wish.[39]

A "back channel" to Seward was opened by the French government when it sent a French agent, John D'Oyley Evans, from Paris with an informal and verbal message from the French foreign minister, Drouyn de Lhuys, and Emperor Napoleon. Calling at the State Department on 17 September 1866, Evans learned that Seward was confined to his room by a severe illness. He informed Frederick W. Seward, the assistant secretary, that the French government would "faithfully and fairly adhere to the very letter of the understanding between France and the U.S. in regard to the evacuation of Mexico."[40] He also added that the Empress Carlotta was received by the Emperor Napoleon as a friend and that some of her requests in regard to the aid for the so-called empire in Mexico had been approved.

Press accounts in France and Mexico about the emperor's disposition to change the evacuation schedule, complained Secretary Seward on 8 October, had produced a large popular mistrust of the emperor's sincerity. He emphasized that the State Department continued to insist upon the fulfillment of the letter and spirit of the evacuation of the French forces in Mexico. Clearly, Secretary Seward exhibited nervousness about the French maneuvers, whether reported in the press, or by confidential messengers.[41] And Seward, reading the American newspapers, witnessed the unusual interest of editors in the American foreign policy crisis precipitated by France. Also, because the American diplomatic dispatches were promptly published in the daily press, it seemed American diplomacy was being conducted in the newspapers.

John Bigelow sent an alarming dispatch to Seward, dated 8 November 1866, and explained that the French ruler had decided to delay withdrawal of any troops until spring: at that time he would remove all his troops, but none before that time.[42] Recent successes of Mexican troops, reinforced by American volunteers, required the continued presence of all the French forces. Moreover, the emperor assured Bigelow that he had telegraphed the message to delay troop removal to Bazaine in plain text, not cipher, in order to forestall any rumors about new secret French designs in Mexico. When Bigelow protested that the French government may not have notified President Andrew Johnson of this dangerous change in plans, Napoleon replied that the existence of the new Atlantic cable lessened the threat of communications misunderstandings.[43] Finally, Napoleon III related that he had advised Maximilian to abdicate.[44]

Seward read Bigelow's dispatch with anger and frustration. In addition, the Republican administration had just witnessed defeat in the recent congressional elections. Some of the opponents were planning to attack President Johnson in the Congress. A forceful cable to France might overcome the opposition, or at least lessen its criticism. And promptly releasing the dispatch to the newspapers would demonstrate the administration's resolve.[45]

Seward's stern reply of 23 November (transmitted 24 November), encoded in the Monroe code first used in 1803, was completed a day after receiving Bigelow's dispatch, and the response was scheduled for transmission on the transatlantic cable: Seward thought in accord with the trial cost basis reached with Hunt at the previous August dinner in New York City.[46] Seward said that his message was written by him with the expectation Bigelow would read the dispatch to the Emperor. Because of this, no word was left out for reasons of economy. Also, before transmitting, Seward submitted the message to President Johnson and the cabinet, which met in an unusual session the afternoon of the 23d, and they approved Seward's dispatch without amendment or change.[47] One cabinet member, perhaps the secretary of the interior, commented on the potentially costly expense of sending the cable; however, Seward explained to the president and the cabinet that he had made an arrangement with Mr. Hunt at the Delmonico's dinner whereby he could set the price for any dispatch he chose to send ("I should pay a *quantum meruit* which the department should fix"). Also, Seward testified later, he had directed one of his subordinates to inform Mr. Hunt of the dispatch at the time of the transmission: he had no recollection whether this was done or not.[48] Actually, someone had alerted Hunt to the existence of the cable, and Hunt telegraphed Seward on Sunday, 25 November, that the dispatch had been sent on to Paris on the previous night.[49]

The encoded Seward dispatch, termed a "pungent remonstrance to the French government" by *The New York Herald*, was given at 6 P.M. on 23 November to the manager of the War Department telegraph office, twenty-six-year-old Charles A. Tinker, for transmission.[50] Tinker recalled the original dispatch was written only in figures and that cable office rules required him to spell out the figures in letters and transmit the letters and figures. He immediately sent for another operator to make a copy of the dispatch so that he might return the original to the State Department and still retain one for his files. Tinker began to transmit the dispatch by 6:15, and it was repeated back to his office so that by 12:15 A.M. the process was finished. It was the longest cable dispatch – 3,722 words – he had ever sent. Incidentally, the State Department used the War office for telegraphing until the State Department moved to the Orphan Asylum Building at 14th and S Streets by early 1867. It then established its own telegraph office linked to the main Western Union office through a special arrangement with the Fire Alarm Telegraph Company.[51]

The Seward historic cryptographic document became the first encoded American diplomatic dispatch to use the new Atlantic cable, that extraordinary communications tunnel. A State Department clerk, John H. Haswell, who prepared the cable, recalled much later: "The first cablegram [actually it was the second] sent by the Department was an important one addressed to our minister at Paris. It caused the French to leave Mexico. I was directed by the Secretary to send it in cipher, using the Department's code which had been in vogue since colonial times but seldom used." Despite its age, Haswell wrote, "It was a good one, but entirely unsuited for telegraphic communication. Its cumbersome character, and what was of even more importance, the very great expense entailed by its use impressed me, and turned my attention to an arrangement for cipher communication by telegraph."[52]

Seward's arguments in the cable, formulated like a lawyer's brief, stressed that the emperor had failed to confer with or notify President Johnson regarding modification of the earlier troop withdrawal schedule. Moreover, the evacuation promised for the spring offered no guarantee of fulfillment; and the change in the timetable interfered with ongoing extraordinary efforts of the United States to cooperate with Mexico for pacifying and restoring proper constitutional authority in the southern republic. Seward concluded with the expectation that the emperor would telegraph or mail a satisfactory resolution in reply to this dispatch; moreover, he wrote that President Johnson believed the French expeditionary forces would be completely removed within the eighteen months originally stipulated.[53]

The New York Herald featured the French evacuation story on 29 November with a brief article under the heading "What is the Meaning of that Long Dispatch?" This account reported a telegram had just been received from London that revealed Bigelow had received a long dispatch and that it was related to "some new hitch in the Mexican difficulty." Additional reports in that newspaper on 1 and 2 December repeated the story that the telegraph focused on the French troops in Mexico; and on 7 December, the *Herald* described Seward's testimony before the Senate Committee on Foreign Relations. Moreover, Seward provided the full plain text of his secret dispatch, which was highlighted on page 4 of the New York newspaper. For more than six decades, the Monroe code had provided a modest degree of protection; however, Seward's maneuvers with the Committee, and possibly the *Herald*, greatly lessened communications security and the value of the code.

The *Herald* also applauded the Seward dispatch with an editorial that stated, "It is an improvement upon all his preceding correspondence on this subject since the close of the rebellion. From that day to this last letter he had been engaged in the unpleasant task of dislodging Napoleon from his 'grand idea' of Mexico by protest, and warnings, and special pleadings, and endless demands for explanations of offensive things done, or for things promised and not performed until the country had lost all patience with this temporizing diplomacy.... Had this decisive course been adopted with the collapse of the rebellion, six months thereafter we might have had the Mexican republic reinstated in the Mexican capital." With accuracy, the editor concluded, "As it is, there is something of credit due even to Mr. Seward, for the patience, the diligence, and the tenacity with which he had held to his text, until we may say he has literally scolded Napoleon out of Mexico."[54]

Bigelow did not read the dispatch to the emperor; rather, his calm response to the lengthy cable told of his note of inquiry to the French minister of foreign affairs, who was out of the city. Receiving no answer, Bigelow pressed the issue further with still another inquiry requesting an explanation of the emperor's motives for deferring the partial evacuation of the troops. In an interview on 30 November, the minister of state and government's spokesman in the legislature, M. Eugene Rouher, told Bigelow the transport vessels were ready and waiting at Vera Cruz and that commanders expected to

The Seward encrypted cable began as follows:

Washington,
November twenty-third, eighteen sixty-six

John Bigelow, Esquire,
United States Minister, Paris.

Sir. — Your dispatch, number three eighty-four, 384, in regard to six twenty-eight, 628, six fifty-one, 651, fourteen hundred four, 1404, fifteen fifty-one, 1551, is received. . . .[55]

have the force returned to France by March, at the latest. Rouher's prediction proved quite accurate.[56] Bigelow also used the cable to reply in code to Seward that there would be collective repatriation in March and that the French government desired friendly relations with the United States. The minister also informed Seward that his reply from Paris cost over 9,160 francs ($1,833).[57]

Seward's confidential dispatch to Bigelow contained more than thirty-five transmission errors; some phrases were mistakenly repeated twice in the cablegram. Many of these errors occurred during the rewrite process when the cable clerk substituted words for the numbers; thus, for example, "1424" was incorrectly sent as "fourteen twenty six." Seward's original plaintext message of 780 words, when encoded, became 1,237 number groups with 88 additional code symbols, such as a cross and an arrow, spelled out. These groups and symbols plus the address were rendered into 3,722 words for transmission.[58]

* * * * *

During December, Charles A. Keefer, a cipher clerk for General Philip Sheridan in New Orleans, would provide invaluable information regarding the French withdrawal from Mexico. This young man was one of twenty Union operators who came to the United States from Canada and the other northern provinces.[59] Almost certainly, Keefer was the first in the United States service to use communications intelligence in peacetime. In mid-December, he wrote to General Ulysses S. Grant that he had happened to be in the New Orleans telegraph office on 9 December when a message from Napoleon to General Castelnau in Mexico was being transmitted via the French consulate in New Orleans. He copied the message, translated it, and gave it to General Sheridan, who in turn sent it to Grant.

Keefer also copied an encrypted cable message to Napoleon, dated 3 December, Mexico, and could not decipher it. Hopefully, Keefer wrote, the 373-cable-word message might be published in a French newspaper, and then the American consul or minister could forward a copy to him so he could work out the key in order that he could decrypt future messages between Napoleon and Maximilian. Keefer urged General Grant not to mention the cipher clerk's name in this matter because the telegraph lines were in the control of Southern men, and if they suspected his intentions they would not allow him to come any place where he could hear the instrument "clicking."[60] It is likely Keefer never received the plain text of the encrypted message and therefore could not work out the key; however, this message, from Marshal Bazaine and General Castelnau, was published in 1930 in a biography of General Castelnau.[61]It told of Maximilian's desire to stay in Mexico; in addition, the two French officers wrote that since the evacuation was to be completed in March, it was urgent for the transports to arrive. Would it be possible, they asked, for the French officers and soldiers attached to the Mexican Corps to have the option of returning?

Keefer wrote to Seward directly in early January, telling him the New Orleans newspapers were printing a telegraphic synopsis of the 3 December Bazaine-Castelnau dispatch to Napoleon and requested the secretary to send him a plaintext copy so that he could work out the key to the encrypted intercept he held. He also reported he had intercepted a dispatch from a reporter for *The New York Herald*, sent from New Orleans to the editor, James Bennett. The reporter's dispatch, datelined from Paris, described the fact that the War Cabinet in Vienna had told the Austrian commander of the corvette *Dandelo* at Vera Crux to remain there until further orders, and also that Napoleon knew this. Keefer emphasized the dispatch never came from Paris at all but originated in New Orleans, and the writer told Bennett to publish it as European news from Paris.

General Sheridan found Keefer's aggressive practices of great value, and he rewarded the young man with a cash prize of $1,600 for managing a secret telegraph line, working out the cipher duplicate messages from Napoleon and the Europeans involving Maximilian and others in Mexico, and counteracting the machinations of a secret society in New Orleans and in the South. However, despite Sheridan's statement, there is no evidence in the remaining historical records that Keefer successfully decrypted the French dispatches.[62]

Keefer's secret intelligence work continued with a dispatch to Seward on 11 January: he included the text of a forty-nine-word cable message in French, sent in the clear, from Napoleon in Paris to General Castelnau, dated 10 January. The Emperor cabled as follows: "Received your despatch of the 9th December. Do not compel the Emperor to abdicate, but do not delay the departure of the troops; bring back all those who will not remain there. Most of the fleet has left."[63] Keefer enclosed the complete cable text, transmitted via the French consul in New Orleans, and suggested that it gave a clue to Napoleon's policy for Mexico.

Keefer's final letter one week later to Seward, who was apparently troubled by Keefer's intercept practices, was an apology. The chastened cipher clerk explained his only motive in sending the previous information was to be of service to the government: "I did not exactly consider myself as playing the part of a spy but on the contrary I considered it my duty as cipher operator. . . . to send you copies of the despatches concerning Maximilian."[64] Continuing his letter of justification, Keefer wrote that he realized the secretary of war had removed all restrictions on telegraphic correspondence the previous April; however, Keefer thought the current affairs in Mexico "would warrant me" in telling you of the policy Napoleon intended to pursue towards Maximilian.

Keefer's final request to Seward: please do not mention my name regarding this matter since it would harm my prospects as a telegraph operator on the Southern lines. And this melancholy supplication concluded the first peacetime communications intelligence effort. Apparently, Keefer did not realize that "Gentlemen do not read each other's mail."

* * * * *

Earlier State Department monthly bills in 1866 for using the domestic telegraph lines were modest: for example, those received for September that, with an eight percent discount, amounted to $73.79; for October, $76.34.[65] The November telegraph bill amounted to $46.94. And then came the astonishing charges for the 23 November cable to Bigelow – $19,540.50. This cost together with other cables sent in November added up to $24,996.12, an amount equal to the yearly salary of the president of the

United States and three times more than that paid the secretary of state.[66] Secretary Seward was unwilling and unable to pay the cable charges.

At the request of William Seward, Cyrus Field, the creative manager of the New York, Newfoundland and London Telegraph Company, met with Seward in Washington to discuss the $25,000 bill.[67] Wilson Hunt accompanied Field. In many ways it was a delicate mission, for the company desperately wanted the government's business, Seward's good will, and the money. Field did not forget that future cable projects might require American governmental support. During the hour-long visit in the secretary's office, Seward complained that whereas he wrote a dispatch of only 780 words in plain text, and had William Hunter, second assistant secretary of the State Department, put the message in code, the charges were for 3,722 words.[68] Field carefully replied that the message came to the telegraph office in code, and it was transmitted exactly as submitted; moreover, he added, Seward would have considered it a "great piece of impertinence on our part if we had asked him" to change the dispatch. Besides, Field added, the company charged him no more than it charged other governments.[69]

Embarrassed and without sufficient funds, Seward asked Field to accept a partial payment of between $5,000 and $6,000, based on the number of words in the original message; if Field approved, the company would eventually be paid in full, and the department would continue using the cable frequently.[70] Seward explained that Congress had not appropriated sufficient funds that would enable him to pay this account. Field then questioned him about the wisdom of using a cipher that had been in use since the formation of the nation. Seward quickly replied that a new economical cipher would replace the old one. In Field's judgment, it was evident Seward had made a great blunder, that when he ordered the dispatch to be put in cipher, he did not realize it would amount to such a large expense. Hunt explained that they were not authorized to accept this $5,000 compromise because his company had already paid the money to the other companies and that at the end of every month, the account was made up. Western Union then took out its money and paid the balance over to the New York, Newfoundland, and London Company, which took out its share. The balance was remitted to London.[71] After a few more minutes of conversation, the secretary finally stated again he would not pay the bill. However, he invited the gentlemen to dine with him.[72]

Somebody leaked the news on the Seward-Field-Hunt private conference to *The New York Herald*, for on 27 December the editor reported inaccurately that the cable company charged $25,000 for the 23 November Seward dispatch and that Seward, not having sufficient funds, paid only $5,000 on it. And then the newsman added with sarcasm: "The United States government must be in a very bad way. All our cable despatches which we have received since the opening of thc line were paid for in gold at the other side of the Atlantic, without any reservation or deduction, and we never made any demand for abatement or delay in the payment." The editor concluded, "It is a shame for the United States government not to be able to pay its telegraph bills as promptly as a New York newspaper."

That same day, Hunt and Field hastily composed a telegram of apology to Seward, explaining that upon their return from Washington, they had reported the results of their Seward interview to the directors of the Telegraph Company; however, where and how the *Herald* obtained its information they did not know, and they regretted the editorial very much.[73] An equally prompt reply from Seward acknowledged their note and added that he had no doubt the journal obtained its information from a source unknown to them.[74]

Though a nervous Napoleon had been "scolded" out of Mexico when the final French troops left Vera Cruz on March 11, the diplomacy between Seward and the New York cable company about the unpaid charges totaling $24,935.75 for the three November cipher messages continued to embarrass both parties. However, the State Department continued to use the cable: in December, for messages to Paris, Alexandria, London, and Liverpool with one message in code, and five messages in plaintext at a total cost of $743.50. Three messages in January to London and Copenhagen, two in code and one in plain text, totaled $615; only one message, to Nice, for $77.25 was sent in February. Two messages, one in code, one in plain text to London in March, at a cost of $1,157.50, were transmitted.[75] The charges for all these cables were paid in gold by the department in early May when Leonard Whitney presented the bill to Seward in person; however, the bill for the three November code cables remained unpaid. Seward told Whitney that Field and Hunt knew the reasons for his refusal.[76]

Another unique cable dispute involving Seward began on Monday, 25 March 1867, with the transmission of an encrypted 1,833-word (the cable company called them "words"; however, they were cipher characters) cable from the Russian minister, Edouard de Stoecki, to St. Petersburg. The dispatch began:

```
t5e5lydzs7x212kvzzkgte74z6x0ykj8vwz747ng
20p5jg1gwy3x7zt8e8t2dkg8yfzlk3ytde69ssp5
0yt4krr1l0kkftx122g2k5n3etgfnjtrfj1yx6k1zdl
gw3pn55
```

and continued for more than forty-nine lines of encryption. This message is the first encrypted cable ever sent by a foreign minister over State Department lines. It was transmitted through the newly organized State Department telegraph office to Prince Aleksandr Gorchakov, vice chancellor of the Russian Empire, in St. Petersburg at a cost of $9,886.50.[77]

The lengthy cable by the dean of the diplomatic corps in Washington and Seward's friend, contained, encrypted in French, the basic treaty conditions for the purchase of Russian America for $7 million. Stoeckl closed the cable with a firm note of economy and extreme urgency: "I send this telegram at the request of Seward who pays for it and who said to me that he has met with great opposition in the Cabinet because of the sum agreed on and that for the affair to succeed it will be necessary to make haste and to have the treaty confirmed by the Senate which is to sit for two weeks longer. If I receive reply within six days the treaty can be signed and confirmed next week by the Senate."[78]

The Russian government promptly replied to Stoecki with qualified approval; $200,000 had to be added to the price in order to cover any claims by the Russian-American Company. Seward, anxious to acquire this vast territory, agreed and quickly prepared the necessary documents. Final negotiations for the purchase of Alaska, which Seward considered his greatest achievement as secretary of state, concluded at 4 A.M. on 30 March with the signing at the State Department office. According to one account, Seward, hoping to win over the recalcitrant chairman of the Senate Foreign Relations Committee, Charles Sumner, invited him to the early morning signing ceremony; however, Sumner went to Seward's residence by mistake and missed the function.[79]

Sumner supported the expansionist treaty, and the Senate advised ratification on 9 April by an overwhelming vote of 37 to 2. After approval by the president, and the exchange of ratifications, the treaty was proclaimed on 20 June 1867, and the Johnson administration took possession in

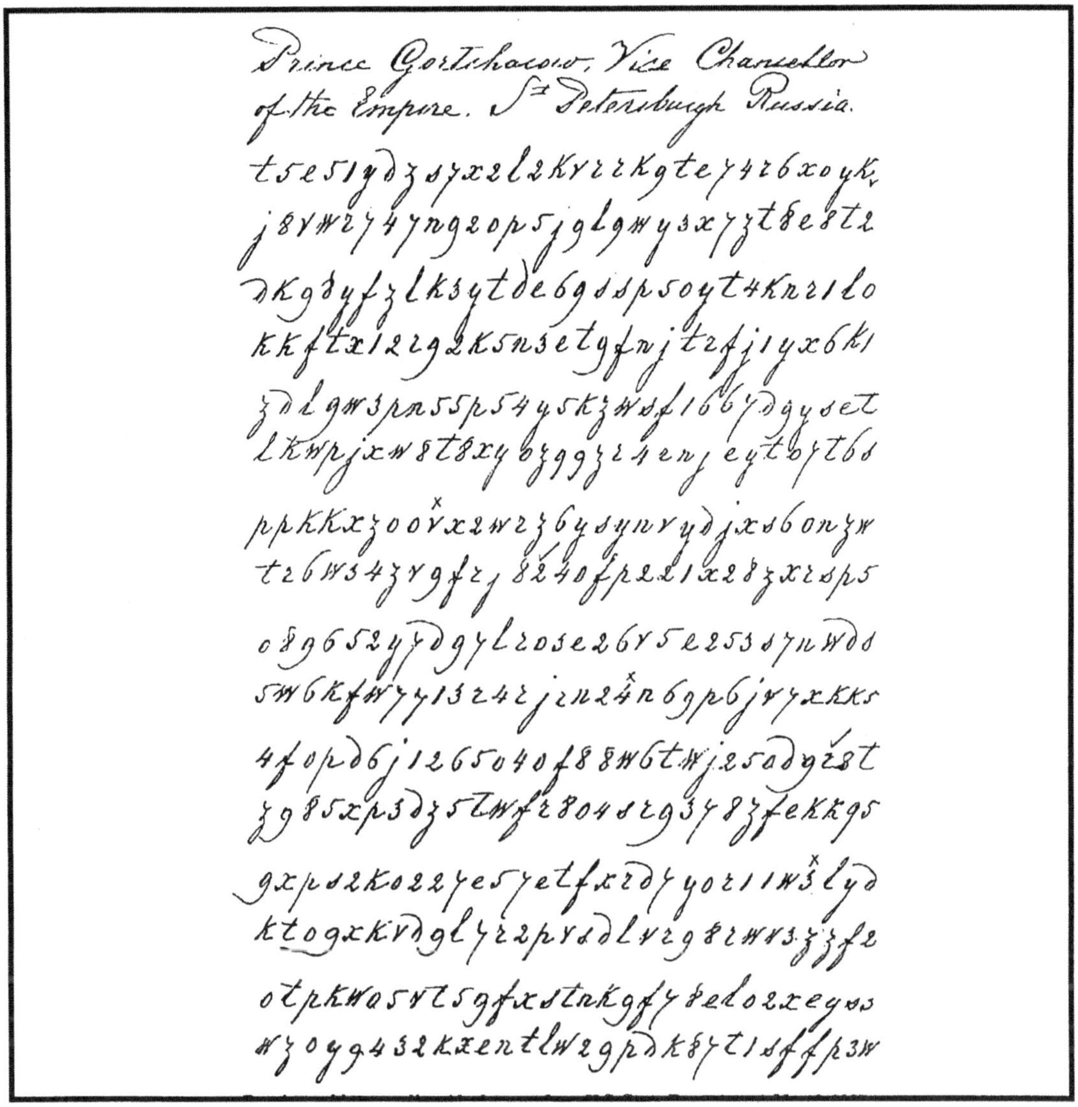
Prince Gortchacow, Vice Chancellor of the Empire. St Petersburgh Russia.

Russian cable regarding Alaska sent from U.S. State Department, March 1867

mid-October. However, a reluctant House of Representatives delayed appropriating the necessary funds until 14 July 1868. The full purchase price in gold was paid to Stoeckl; however, he sent only $7,035,000 to Russia. Apparently, lobbyist Robert J. Walker received around $26,000; editor John Forney, between $4,000 and $30,000; and ten congressmen were paid sums ranging from $8,000 to $10,000. Stoeckl told Walker that cables to St. Petersburg and all the other costs had been very expensive.[80]

As noted above, Whitney's visit to Seward on 3 May resulted in a partial payment of cable charges. However, now almost $10,000 for the Russian encrypted cable originally charged to the Russian legation was transferred to the American account at the order of Stoeckl. In addition, two cables from Seward to Adams on 15 and 23 May, sent in the Monroe code, added another $7,300 to the unpaid account, bringing the total to over $42,000. The troublesome account also increased Hunt's and Field's financial anxieties by late May. Hunt telegraphed Seward, stating he and Field were going to Washington and asking if it would be convenient for them to visit the secretary. An adamant and adroit Seward promptly replied he would be delighted to see them socially at any time; however, he would not hold any interview concerning the cable telegrams. He also cabled his minister in France, John A. Dix, and Charles Francis Adams in London to "use the cable no more in cipher or writing. It will not be used here."[81]

A disappointed Hunt, still financially sensitive to Seward's power, quickly replied by letter on 1 June to Seward and recounted the previous tariff schedule and the dinner conversation at Delmonico's, including Hunt's understanding that Seward would write to him about reducing the cable charges; however, Hunt again explained, no letter from Seward had arrived. During November, he continued, the State Department dispatches were promptly transmitted but never paid. Instead, the New York Company, which would have kept less than one third of the amount, remitted two thirds of the bill out of its own funds to London for payment. Further construction expenses by the Newfoundland Company for two new landlines in Newfoundland and a contract for a sea cable to be laid from Newfoundland to the French island of St. Pierre, and thence to Sydney, were pressing the company treasury. Hunt concluded cautiously, "Although the company are greatly in want of money, they would not press their claim at this time if it be inconvenient or embarrassing to the Government. But the company have a greater trouble, and one that is exceedingly embarrassing, that is a refusal on the part of the Government, after having used the telegraph, and we having assumed and paid two-thirds for the Government, to acknowledge the debt."[82] Hunt did not mention the bill for the Russian cable.

Always a tough negotiator, Seward sent a two-sentence reply: "I have received and attentively read your letter of the 1st instant. I am, dear sir, Your obedient servant."[83] One week later, Leonard Whitney, cashier for the telegraph company, asked George Baker, the department accounting clerk, if he could collect for the May cable messages and received a prompt "No."[84]

Seward's unhappiness with the cable costs for transmitting dispatches masked by the Monroe code brought into existence the first new State Department code in fifty years. This extremely awkward code, devised for economy, was based upon the letters of the alphabet. The twenty-three words most frequently used in dispatches were assigned one letter of the alphabet. For example, "a" was The; "b" was It; "c" was Have, and so on. "W" was not used for the code (though it was in cipher) because European telegraph operators were not familiar with this letter. The next 624 most frequently used words were encoded by two letters of the alphabet: for example, "ak" for Those; "al" for Who; and "az" for Such. Three letters were used for the remainder of the diplomatic vocabulary, and a fourth letter could be added for plurals, participles, and genitives.

On 19 August 1867, a copy of the new code was sent to John A. Dix, minister to France, and to Cassius Clay, minister to Russia, and to other ministers.[85] For security purposes, Seward asked that the code be used with discretion and also that the minister should have a small box made that could be fastened with a lock, the key to which should be kept by the head of the legation.

This novel code, which delighted the thrifty Seward, was used between August 1867 and 1876 but proved to be a disaster because European and American telegraphers often merged code groups, and dispatches were frequently unread until mailed copies reached the State Department weeks later. Indeed, the first encoded message received at the department from the American minister in Turkey formed a long string of connected letters and remained a conundrum until finally decrypted by an assistant clerk after days of puzzlement.

Similar messages came from Paris and one from Vienna; the latter one was never decoded.[86] Seward's battle with the cable company resulted in this supposedly thrifty but flawed encryption system.[87]

The New York Company and Hunt did not contact Seward again until the company completed its new tariff. On 1 December 1867, a new schedule lowered rates one half: messages from New York to London would cost $25 in gold for ten words, each word containing no more than five letters. Five words, not in excess of twenty-five letters, for address, date and signature would be free. Moreover, messages in code carried no extra charges: ciphers were to be charged $25 for the first ten letters. This next tariff schedule change may have pleased Seward when it stated, "Government using a code shall pay for the number of words contained in the dispatch before it be translated into code, provided that the code be so constructed that not more than four letters or four numerals be used to constitute a word." Apparently, although it was not stated, the cable company would have to accept the government's word total for encrypted messages. Hunt added that for governments represented in Washington and the newsmen in Europe and the United States, the new schedule took effect with the date of his letter, 20 November 1867.[88]

In notifying Seward of these modifications, Hunt politely renewed his request for payment of outstanding charges, which, including the Russian cable and other State Department cables sent in May and October, brought the total charges to $42,289. Moreover, the New York Company had already paid out $28,923.46 to Western Union and the European companies for these State Department dispatches. By way of compromise, Hunt offered to cancel tolls of approximately $13,000, that is, New York's charges for transmitting the government dispatches over their line from Port Hood, Cape Breton, to Heart's Content, on the eastern shore of Newfoundland, if the government would pay the $28,923.46. Appealing to Seward's patriotism, Hunt noted nine tenths of the New York stock was owned by citizens of the United States.[89]

Seward's reply, written exactly one year after preparing the lengthy encrypted cable to Bigelow, praised the tariff reductions and noted the charges would be advantageous for his department since a new code was being used, with every word in the English language expressed in an average of fewer than three letters. He regretted that no reduction was made of the previous charges in the department's account; also, he added firmly that the department was not accountable for the Prince Gorchakov cable since that dispatch was not signed or ordered by him.[90]

Two months later, Hunt again apologetically wrote and explained that Abram S. Hewitt had gone to London as the special commissioner of the New York Company to negotiate with the Atlantic Telegraph Company for settling the department's account, "this vexatious business." Hewitt sought unsuccessfully to have the monies paid to Atlantic Telegraph refunded to the New York Company so that a compromise could be reached with Seward. In addition, Hunt wrote that after Seward's 23 November letter, he contacted Baron Stoecki regarding the Russian cable and was told that Seward had agreed to pay the cost of the cable to St. Petersburg and that Stoecki would pay for the reply. Politely, a weary but determined Hunt asked Seward to come to an understanding with Stoeckl so the proper party would pay the bill.[91]

Once again, Seward replied crisply: nothing in Hunt's letter modified the views of the department as expressed earlier to Field and Hunt. Nor would the department discuss the claims of the Atlantic Telegraph Company upon the Russian minister.[92]

The tedious exchange of polite letters continued into January 1868. This time, Peter Cooper, Moses Taylor, Marshall Roberts, Cyrus Field, and Wilson Hunt, all directors of the New York Company, prepared a joint dispatch to Seward. They carefully corrected Seward's assertion by stating the Atlantic Telegraph Company had no claim against the department; rather, Newfoundland had already paid all the tolls to the other companies, including the charges for the Russian dispatch. They continued: "We are at a loss to perceive in what respect we have erred, or why we should merit such punishment" and as businessmen, they protested Seward's refusal to fulfill the department's financial responsibilities.[93]

Instead of his usual curt reply, Seward recalled that the department's views were expressed on Hunt's and Field's first visit to the department. Now, however, he added new information to the narrative by explaining he had written Charles Francis Adams, American minister to England, soon after seeing Hunt and Field. He had explained to Adams that the legation and the department had no funds to meet the high costs, especially for cipher messages. Adams was also informed that a prominent New York proprietor of the telegraph, learning of Seward's evaluation, invited the department to use the telegraph with cipher "leaving that question of compensation to be determined by the Department itself....[94] With this understanding, wrote Seward, the telegraph was used with cipher for special occasions. However, charges were billed on the basis of the regular tariff. In addition, use of the "long-used cypher" increased about threefold the number of words transmitted. Because of this, Adams was told to use the cable only for emergencies of very great urgency. Finally, Seward wrote to the directors, the rejected accounts charged under the original tariff were extortionate and objectionable because the charges did not conform to Mr. Hunt's promises.

One week later, 12 February, Hunt replied to Seward's charges by recalling the Delmonico dinner and his understanding that the secretary would write to the directors with a request to lower the charges on the cable tariff. No letter came. Next, Hunt added, a message from the department to Bigelow in Paris was sent to the telegraph office. Also, Hunt explained for the first time, Seward had sent a letter, dated 24 November, to Hunt, and it was received by him the next day, Sunday morning. Seward asked that the encrypted dispatch be sent forward without delay. Upon inquiry, Hunt learned the dispatch had been transmitted the previous evening; therefore, he supposed the Seward's letter referred only to prepayment of the dispatch since regulations in Europe and America required payment in gold before transmission. In fact, the only exception to this rule, explained Hunt, was for the United States.

Moreover, Hunt wrote, he never had the power to change the tariff because it had been established by the three companies, and only they could make modifications. Seward's complaint about the cipher cable costs was not consistent with his actions, Hunt explained further, because the department sent twelve dispatches between November and April and paid for them without complaint. Now, however, the department had sent four messages in May under the same tariff and refused to pay these charges, which Seward called exorbitant. Hunt closed the letter with a conciliatory paragraph in which he noted the ocean telegraph was greatly indebted to Seward for his early assistance and eloquent appeals for support in the U.S. Senate. His contributions secured a new dimension for the preservation of peace and the progress of civilization.[95]

Field and the other directors continued to worry about damaging the cable company's relations with the United States government. On 10 March, at the Telegraphic Banquet in the Palace Hotel in London, Cyrus Field and his guests, celebrating the telegraph company, sent a special greeting to President Andrew Johnson with the hope that "the Telegraphic Union between England and America may never be interrupted nor their friendship broken."[96] Their worry focused on American friendship.

Reluctantly, over seven months later and aware that a presidential election would soon occur, Simon Stevens, a lawyer for the cable company, wrote Seward and suggested the entire matter be referred to the attorney general for his opinion, which the company was prepared to accept as final. In addition, the counsel enclosed a letter from Richard Lathers, who replied to Hunt's request for his understanding of the discussion after the Delmonico dinner. Lathers's letter supported Hunt's recollections exactly.[97]

Two years after the famous Seward-Bigelow cable was transmitted and with only three months remaining as secretary, Seward wrote his last letter to the cable company and explained in one sentence that he had no authority to make, nor the attorney general to entertain, an adjudication of the claim.[98]

Appealing one year later to the new secretary of state, Hamilton Fish (and exactly three years after the Seward-Bigelow cable), the telegraph company directors recounted in several pages the history of the unpaid department cable charges, which added up to $32,240.75 for the three dispatches in November 1866 and two more in May 1867. The directors made no mention that the Russian cable charges of almost $10,000 had been paid by the Russian legation in Washington, D.C.[99] Hamilton Fish's reply repeated Seward's statement two years earlier that the department would pay for the cables based on the number of words, not ciphers. In addition, he also endorsed his predecessor's determination regarding referral to the attorney general and refused to reverse this decision.[100]

MEMORANDUM OF ACCOUNT WITH DEPARTMENT OF STATE

Department of State of the U.S.,

TO THE NEW YORK, NEWFOUNDLAND AND LONDON TELEGRAPH COMPANY, DR.

MESSAGES RECEIVED	FROM WHOM	TO WHOM	DESTINATION	NO. OF WORDS		AMOUNT COIN	DATE OF PAYMENT	AMOUNT PAID, COIN	AMOUNT UNPAID, COIN
1866							1867		
Nov. 10 ...	Seward	Bigelow	Paris	23	E	$60 37	May 4	$60 37	
" 24 ...	Seward	Bigelow	Paris	3722	C	19,540 50			$19,54050
" 29 ...	Seward	Adams	London	280	C	1,400 00			1,400 00
" 30 ...	Seward	Bigelow	Paris	761	C	3,995 25			3,995 25
Dec. 1 ...	Seward	Bigelow	Paris	74	C	388 50	" "	388 50	
" 3 ...	Seward	Hale	Alexandria	36	E	112 50	" "	112 50	
" 3 ...	Seward	Adams	London	16	E	50 00	" "	50 00	
" 11 ...	Seward	Dudley	Liverpool	26	E	65 00	" "	65 00	
" 17 ...	Seward	Dudley	Liverpool	30	E	75 00	" "	75 00	
" 28 ...	Seward	Dix	Paris	19	E	52 50	" "	52 50	
1867									
Jan. 10 ...	Seward	Stevans	London	30	E	75 00	" "	75 00	
" 12 ...	Seward	Yeaman	Copenhagen	63	C	330 00	" "	330 00	
" 29 ...	Seward	Adams	London	42	C	210 00	" "	210 00	
Feb. 5 ...	Seward	Aldis	Nice	30	E	77 25	" "	77 25	
March 7 ...	Seward	Adams	London	215	C	1,075 00	" "	1,075 00	
" 25 ...	Seward	Adams	London	33	E	82 50	" "	82 50	
" 25 ...	———	Gortschacoff* ...	St. Petersburgh ..	1833	E	9,886 50	A'g 22, 1868	9,886 50	
May 15 ...	Seward	Adams*	London	575	C	2,975 00			2,975 00
" 23 ...	Seward	Adams*	London	866	C	4,330 00			4,330 00
" 24 ...	Seward	Adams	London	22	E	55 00	J'ne, '68	52 00	
" 24 ...	Seward	Dix	Paris	22	E	56 75	" "	56 75	
July 16 ...	Seward	Adams	London	13	E	50 00	July 20	50 00	
" 22 ...	Seward	Adams	London	14	E	50 00	Sept. 7	50 00	
" 28 ...	Seward	Adams	London	12	C	100 00	" "	100 00	
" 28 ...	Seward	Adams	London	15	E	50 00	" "	50 00	
Sept. 3 ...	Seward	Yeaman	Copenhagen	26	C	137 50	" "	137 50	
" 19 ...	Seward	Adams	London	41	E	102 50	" 23	102 50	
" 19 ...	Seward	Hale	Madrid	14	E	53 50	" 20	53 50	
Oct. 5 ...	Seward	Yeaman	Copenhagen	9	C	104 50	Oct. 14	104 50	
Total						$45,540 62		$13,299 97	$32,240 75

* Transmitted direct by Telegraph from office in Department of State

Cable Company Memorandum of Account with Department of State

Finally, on 25 February 1870, the New York, Newfoundland and London Telegraph Company filed a petition in the United States Court of Claims and requested that the government pay $32,240.75 in gold coin for the cable messages from the Department of State to Paris and London. The petitioner added a patriotic statement that the company directors "at all times have borne true allegiance to the Government of the United States, and have not in any way voluntarily aided, abetted, or given encouragement to the rebellion against the said government."[101]

The "Argument for the Claimant," covering twenty-six pages, submitted on 13 March 1871, to the U.S. Court of Claims for the December term, 1870, reviewed the previous correspondence and depositions taken in the case. Especially notable was Hamilton Fish's agreement that the accounts in the claimant's petition were accurate except for the Russian cable, which the State Department neither authorized nor paid. The claimants agreed with Fish's assertion. The Argument also highlighted the Delmonico dinner and the conversations between Hunt and Seward as stated in the depositions, and also that of Richard Lathers, before coming to the conclusion that there was no evidence for a special agreement, binding upon the claimant, through which the United States government would have the right to send telegrams over its own and connecting lines at rates lower that the customary charges for sending telegrams by private parties. Thorough in gathering data for the Argument, the lawyers for the claimants also emphasized that the appropriations were adequate for payment of the charges: contingent expenses in the diplomatic service, though reduced for 1867, were for the years ending in June as follows:

	For All Missions Abroad	For Foreign Intercourse
1866	$60,000	$80,000
1867	$50,000	$65,000

Citing more than twenty court cases concerning various aspects of the dispute between the cable company and the State Department, the New York, Newfoundland, and London Telegraph Company lawyers concluded that the claimant should recover the $32,240.75 unless "its rights of recovery is [*sic*] defeated by the pretended agreement, alleged to have been made between Mr. Seward and the claimant, previous to sending of said dispatches."[102]

The United States's defense regarding the claim specified the government never agreed to pay for the telegraphic service at the published rates. Rather, wrote Thomas H. Talbot, assistant attorney general, it agreed to pay an amount deemed by the secretary of state to be proper compensation. In his deposition, dated 8 August 1870, Seward thought the sum of $5,600 in gold would be a fair, just, and reasonable compensation for the telegraph services.[103]

The case was heard before the chief justice and judges of the Court of Claims in Washington, D.C., on 26 May 1871. In its "Findings of Fact and Conclusions of Law," the court found that the data presented by the claimants were correct, that the secretary of state had paid charges for twenty-three cables (of which seven were encrypted) at regular rates and that he refused to pay five other cable charges, all of them encrypted. Moreover, the company had paid $21,804.90 in gold coin to the connecting lines and was owed this amount plus $10,435.85 for transmission over its own lines, for the total of $32,240.75.

The Finding also stated that neither previous to, nor at the time of, the sending of the dispatches was there any valid agreement between the secretary of state and the claimant whereby the dispatches were to be transmitted at any other of different rates than those customary rates in force at the time they were transmitted. The secretary had not written a request for rate changes, nor did Wilson Hunt have the power or authority to change or modify the rate schedule established by the joint action of the connecting telegraph companies.[104]

In its conclusions, the court found the secretary had the power to bind the United States for the "fair and reasonable charges" for transmitting the dispatches. The charges were high but not so exorbitant under the circumstances so that the court would be justified in reducing them. And most importantly, there was no "specific or formal contract between the parties," that the conversation between Secretary Seward and Wilson Hunt at Delmonico's in New York was a "mere incidental conversation and without the force and effect of a contract."[105]

The court decided for the claimant in the amount of $32,240.75. The State Department had one victory: payment in gold was not required.[106] Rather, the judgment had to be rendered "in the usual form in dollars and cents, without distinguishing the kind of money in which it shall be paid." Promptly, the New York, Newfoundland and London Telegraph Company's treasurer, Moses Taylor, wrote to the secretary of the treasury requesting that the judgment be immediately paid, or five percent interest be added until paid. He enclosed a certified transcript of the judgment.[107] And finally, on 28 August 1871, almost five years after the Seward-Bigelow cable, the Comptroller's Office paid the full amount in dollars and cents.[108]

Notes

1. John Bigelow, *Retrospections of An Active Life* (New York: The Baker & Taylor Co., 1909), III, 611. Also cf. Beckles Willson, *American Ambassadors to France 1777-1928* (London: J. Murray, 1928), 287-288. Seward went to extreme lengths to continue the charade, and his numerous dispatches concerning French forces in Mexico continued to flow out of the State Department until March 1861.

2. Bigelow, *Retrospections*, 111,612.

3. Bigelow to Seward, Paris, 3 August 1866, in Record Group 59, General Records of the Department of State, Dispatches from U.S. Ministers to France, Microcopy 34, Roll 62, National Archives. Hereafter cited as RG 59, M34,R62,NA.

4. Ibid. Several years later, Hamilton Fish would cable Robert Schenck, American minister in London, in code and urge him to use code in his dispatches because the telegraph office was leaking information to the newspapers in the United States: cf. Fish to Schenck, Washington, D.C., June 16, 1872, Hamilton Fish Papers, Letter Copy Book, 13 March 1871 to 25 November 1872, Library of Congress, hereafter cited as LC.

5. Bigelow to Seward, 3 August 1866, RG 59, M34, R62, NA.

6. Seward to Bigelow, Washington, D.C., 21 August 1866, Record Group 59, Diplomatic Instructions of the Department of State, Microcopy 77, Roll 58, National Archives. Hereafter cited as RG 59, M77, 58, NA.

7. A determined Bigelow replied to Seward's dispatch by again stating the strong possibility that the cipher at Paris and other legations had been violated by the "treasonable affinities of Mr. [William] Dayton's immediate predecessors...." cf. Bigelow to Seward, Paris, 12 October 1866, RG 59, M34, R62, NA.

8. Bigelow, *Retrospections*, 111,627.

9. *The New York Times*, 30 August 1866. Tensions between Johnson and the Congress over Reconstruction issues continued to expand. Several weeks later, General Ulysses S. Grant sent a confidential letter to General Philip Sheridan and reported on these differences and the fear that Johnson may have wanted to declare Congress illegal, unconstitutional, and revolutionary. Cf. Grant to Sheridan, Washington, D.C., 12 October 1866, Sheridan Papers, Box 39, LC.

10. Ibid., 30 August 1866. An editorial that same day emphasized that Johnson's visit and western tour to Chicago would give him "new strength to carry on the war for the Constitution, which he has thus far waged with such signal power and success." Moreover, the editor wrote

that Johnson, Seward, General Ulysses S. Grant, and Admiral David Farragut "did more than any other four men in the country to suppress the great and terrible rebellion of the Southern States."

11. Deposition of Richard Lathers, 23 August 1870, Records of the U. S. Court of Claims, General Jurisdiction Case Files, 1855-1937, Case No.6151, Record Group 123, Box No. 306, National Archives. Hereafter cited as RG 123, B306, NA. Cf. also "Argument for the Claimant" in United States Court of Claims, December Term, A.D. 1870, Filed 13 March 1871, in RG 123, B306, NA.

12. Deposition of Wilson G. Hunt, 8 June 1870, RG 123, B307, NA. In his deposition, Lathers said at first he thought Seward was being facetious, and the conversation began rather jocularly; it then turned serious as Hunt listened carefully to Seward's criticisms.

13. Petition of the New York, Newfoundland & London Telegraph Co. vs. The United States, filed February 25, 1870, Claim No. 6151, Record Group 59, Microcopy 179, Roll 319, 7, NA. Hereafter cited as the Petition, RG 59, M179, R309, NA. Domestic or landline charges by the Western Union Company were at the regular rate for code and cipher messages; however, cable charges were double: cf. Deposition of Charles Tinker, 16 September 1870, RG 123, B307, NA. Code and cipher messages continued to trouble the telecommunications executives and several systems were tried: for example, fees were based upon five characters per word: cf. James M. Herring and Gerald C. Gross, *Telecommunications: Economics and Regulation* (New York and London: McGraw Hill Book Co., 1936), 138-147.

14. In November 1866, a gold dollar equalled about $1.40 in greenback currency: cf. Wesley Clair Mitchell, *Gold, Prices, And Wages under the Greenback Standard* (Berkeley: The University Press, 1908), 302.

15. Deposition of William Seward, 27 July 1870, RG 123, B307, NA. Seward knew that Hunt, Peter Cooper, and Cyrus Fields, all of New York, were principals in the cable company.

16. Ibid.

17. Deposition of Hunt, ibid.

18. Irwin Unger, *The Greenback Era: A Social and Political History of American Finances*, 1865-1879 (Princeton: Princeton University Press, 1964), 16.

19. Deposition of Seward, RG 123, B307, NA.

20. All three depositions, by Hunt, Lathers, and Seward, mention that the secretary would write to the New York telegraph company and offer suggestions for lower rates.

21. Congressional legislation, approved 31 January 1862, authorized the president of the United States to take military possession of the telegraph and railroad lines in the nation. However, in his deposition of 16 September 1870, Charles Tinker, the War Department telegrapher, testified the charges for the government messages sent over the Western Union lines were the same as those for private individuals. The only exceptions were messages sent over the Pacific telegraph lines: these lines were subsidy lines, and the government rate was lower than that fixed for private concerns.

22. Seward's most recent biographer wrote: "Seward was an agitator, a politician, and a statesman, all in one. His irresistible impulse to pose and explain and appear all-wise and all-important earned for him a reputation for insincerity and egotism. A perfectly fair-minded contemporary gave this answer to a question: 'I did not regard Seward as exactly insincere; we generally knew at what hole he would go in, but we never felt quite sure as to where he would come out.' It is a paradox that precisely explains the paradoxical Seward." Cf. John M. Taylor, *William Henry Seward: Lincoln's Right Hand* (New York: Harper-Collins, 1991), 528.

23. *The New York Times*, 13 September 1866.

24. Ibid., 26 October 1866. Less than one year later, this newspaper reported on the financial success of the cable: that two thirds of the entire outlay spent on the cable on 1866 would be returned in revenue from the first year's operation. Moreover, if one were to add the cost of the cable of 1865, the return would be about thirty percent. Thus, rates should be lowered to a moderate scale, wrote the editor, for then the press dispatches could be doubled in length, and "more than doubled in value of their contents." Ibid., 11 July

1867. The source for these cable revenues was not noted by the editor. George Saward, the secretary and general manager for the Atlantic Telegraph Company, stated that the first year's operation of the cable provided a return of only two percent on the capital investment: cf. Saward to Hunt, 25 November 1867, in RG 59, M179, R27, NA.

25. The Petition, RG 59, M179, R319, 10-12, NA. As specified in the tariff, all figures in the transmission had to be expressed in words, and charged accordingly. Seward to Bigelow, Washington, D.C., 10 November 1866 in RG 59, M77, R58, NA.

26. Seward to Bigelow, Washington, D.C., 10 November 1866 in RG 59, M77, R58, NA.

27. Deposition of Wilson Hunt, 8 June 1870, RG 123, B307, NA. Also the Petition, RG 59, M179, R319, 15, NA. According to the deposition of Charles Tinker, 16 September 1870, RG 123, B307, NA, the cable company required that the payment in gold for the dispatches should be remitted weekly with a copy of the dispatches. However, in practice, only the dispatches were sent forward, and the Western Union Company billed the particular government office that was specified on the copy of the dispatch.

28. Charles Bright, *Submarine Telegraphs: their History, Construction, and Working* (London: Crosby Lockwood & Son, 1898), 80. As Senator, Seward had urged the federal government to back Field's first cable laying venture in 1858.

29. *The New York Times*, 16 November 1866.

30. A lengthy editorial in *The New York Times*, 24 November 1866, predicted that in the event of war with the United States, England would monopolize the cable. The editor noted the cable was owned substantially by English capitalists who controlled the tariff and managed the work, and they also owned the only steamer ever capable of laying a Atlantic cable. And with accurate insight, though with some exaggeration, he added, "The absolute monopoly of telegraphic communication with this Continent, involves a power more vast and terrible than has ever been enjoyed by any nation in the world." Messages passing between America and the Continent would be subject to inspection by the British government. And "We might as well expect the English navy to remain neutral, in case of war with us, as the English Atlantic Cable." Congress, the editor concluded, should appoint a commission to study the question further in order to remedy the evil as soon as possible.

31. Ibid., 16 November 1866.

32. Ibid.

33. A notable Republican editor called the crisis the "greatest diplomatic difficulty our Government has had for two years," cf. ibid., 21 November 1866.

34. Romulus M. Saunders to James Buchanan, Madrid, 17 November 1848, in Record Group 59, Dispatches from United States Ministers to Spain, 1792-1906, Microcopy 31, Roll 3, NA.

35. As quoted in Bernard Porter, *Plots and Paranoia: A History of Political Espionage in Britain* 1790-1988 (London: Unwin Hyman, 1989), 78.

36. John H. Haswell, "Secret Writing," *The Century Illustrated Monthly Magazine*, 85 (November 1912), 89.

37. On 5 April 1866, *Le Moniteur*, Napoleon's official newspaper, noted that French troops would withdraw on these dates. Cf. Frederic Bancroft, *The Life of William H. Seward* (New York & London: Harper & Brother, 1900), 2:438.

38. John Bigelow to William Seward, Paris, 31 May 1866, as reprinted in *The New York Times*, 7 December 1866. Much of the correspondence between Seward and Bigelow, often confidential, regarding the French status in Mexico is reprinted in this issue of the newspaper. In a 16 May letter to Seward, also reprinted, Bigelow quoted from the 15 May issue of the semiofficial newspaper *La France,* to the effect that the embarkation of Austrian volunteer troops from Mexico had been countermanded, the enlisted men were discharged, and the majority of these troops joined Maximilian's forces. The Mexican crisis, including confidential dispatches, fascinated the press, troubled the Congress, and profoundly worried Seward.

39. *The New York Times*, 30 August 1866. Two weeks earlier, John Hay, charge d'affaires in France, wrote to Seward and reported that the French minister promised "the plan heretofore

determined upon by the Emperor's Government will be executed in the way we announced." Cf. John Hay to William Seward, Paris, 17 August 1866, as reprinted in ibid., 7 December 1866. The New York newspaper enthusiastically supported Seward's foreign policies regarding Mexico. Recalling the Monroe Doctrine, the *Times's* editor wrote: "This country is directly interested in this question and the people will insist that its interests shall be protected. Neither France nor any other European Power can be allowed to gain such a foothold on this continent as the establishment of an empire in Mexico, under her protection, would give them. It would be a perpetual menace to our own security." Ibid., I September 1866.

40. Record Group 59, Confidential Memorandum, Department of State, 17 September 1866 in Administrative Records of the Department of State, Reports of Clerks & Bureau Officers, Entry 311, Volume 4, NA.

41. William Seward to John Bigelow, Washington, D.C., 8 October 1866, as reprinted in *The New York Times*, 7 December 1866.

42. Just a month before, Seward had written to Bigelow that the U.S. "relies with implicit confidence upon the fulfillment of the Emperor's engagement at least to the letter, and it has even expected that, overlooking the letter, it would be fulfilled with an earnestness of spirit which would hasten instead of retard the evacuation of the French forces in Mexico." Seward to Bigelow, Washington, D.C., 8 October 1866 in *The New York Times*, 7 December 1866.

43. John Bigelow to William Seward, Paris, 8 November 1866, in RG 59, M34, R64, NA.

44. Bigelow to Seward, Paris, 8 November 1866, as reprinted in *The New York Times*, 7 December 1866.

45. Glyndon G. Van Deusen, *William Henry Seward* (New York: Oxford University Press, 1967), 494-495.

46. At least one principal European power was using a code similar to the Monroe Code: cf. Haswell, "Secret Writing," *Monthly Magazine* 85 (November 1912), 88. Haswell does not specify the particular nation; however, he probably had reference to France, which was using a code with similar Arabic numerals (2209 613 562 273 15 2214 etc.) and which Charles Keefer, General Sheridan's cipher clerk, intercepted in New Orleans in December 1866: cf. Seward Papers, Microcopy, Roll 98, Library of Congress. Hereafter cited as Seward Papers, R98, LC.

47. *The New York Times*, 24 November 1866, noted the session as an "extraordinary convening of that body" and speculated that the discussions focused on the withdrawal of French troops from Mexico.

48. Deposition of William Seward, 27 July 1870, RG 123, B307, NA. When asked if he had written to Mr. Hunt or the New York Telegraph Company regarding his suggestions for lower cable rates, Seward replied he had not done so, "that it would be inexpedient and unbecoming to make such explanations."

49. Hunt to Seward, New York, 25 November 1866, in RG 59, M179, R246, NA.

50. *New York Herald*, 7 December 1866.

51. Deposition of Charles A. Tinker, 16 September 1870 in RG 123, B307, NA.

52. John Haswell to John Sherman, Washington, D.C., 20 January 1898. Photocopy in author's possession: the original letter is in the possession of Mrs. Lester Thayer, Albany, New York.

53. General Philip Sheridan and 30,000 troops just north of the Rio Grande River added emphasis to Seward's message: Taylor, *William Henry Seward*, 269.

54. *The New York Herald*, 7 December 1866.

55. The plain text of the encrypted cable from Seward to Bigelow, Washington, 23 November 1866, may be found in RG 59, M77, R58, NA.

56. Bigelow to Seward, Paris, 30 November 1866, Seward Papers, R98, LC. Also cf. Bigelow, *Retrospections*, III, 622-626, for Bigelow's evaluation of the French situation: that delaying the evacuation was merely an "abbreviation rather than a prolongation of her occupation of the Republic of Mexico."

57. Bigelow to Seward, Paris, 3 December 1866, RG 59, M34, R64, NA. This letter is also reprinted in House Executive Documents, 1:1,39th Congress, Second Session, Serial 1281. Seward honored Bigelow's draft for 9,164 francs, 75 centimes: cf. Seward to Dix, Washington, D.C., 28

December 1866, RG 59, M77, R58, NA. Bigelow's comments on Seward's political maneuvering with Congress regarding Mexico were reprinted in Beckles Willson, *American Ambassadors to France*, 1777-1927 (London: J. Murray, 1928), 287-288. Apparently, Seward went to extreme lengths to continue the maneuvers because his numerous dispatches concerning French forces in Mexico continued to reflect heightened anxiety until the actual troop removal was completed in March 1867.

58. William Seward to John Bigelow, Washington, 23 November 1866, in Record Group 84, Instructions to the United States Legation at Paris, Ci.1, NA. The letter book copy of the dispatch may be found in Diplomatic Instructions of the Department of State, 1801-1 906, RG 59, M77, R58, NA. At 3:10 p.m. on 25 November 1866, the U.S. military telegraph office in Washington received Hunt's telegram to Seward from New York and reported that the dispatch had been sent to Paris the previous night: RG 59, M34, R64, NA. According to *The New York Herald*, 15 December 1866, Bigelow received the first sheet of the encoded dispatch on Monday morning, 26 November at 7:30, and the last page on Tuesday at 4 a. m.

59. Plum, *The Military Telegraph During the Civil War in the United States*, 2:357.

60. Keefer to Grant, New Orleans, 17 December 1866, Seward Papers, R98, LC.

61. Georges A.M. Girard, *La Vie et les souvenirs du General Castelnau* (Paris, 1930), 117-118. The actual message is reprinted by E.C. Fishel in his fine article, "A Precursor of Modern Communications Intelligence," *NSA Technical Journal*, 3 (July 1958), 13-14.

62. Philip Sheridan Papers, Microcopy, Roll 2, LC. Sheridan wrote out this message sometime after 1871. He explained he had given this amount of money to Keefer on or about 24 December 1866; he added that the memorandum and reports on Keefer's operations were destroyed in the Chicago fire of 1871. Chicago was the headquarters of the Military Division of Missouri, and when the city burned, the headquarters and all of General Sheridan's records were destroyed. In an attempt to reconstruct the record, Sheridan had two clerks in Washington copying everything relating to his campaigns as filed in the War Department, and these copies constitute a large amount of the Sheridan Papers in the Library of Congress. Cf. George A. Forsyth to General Adam Badeau, Chicago, 21 November 1873, in the NHPRC Search Sheets for U.S. Grant, Library of Congress.

63. Keefer to Seward, New Orleans, 11 January 1867, Seward Papers, R99, LC. General Philip Sheridan, in his book, mistakenly stated the dispatch was received in cipher and translated by the telegraph operator [Keefer], "who long before had mastered the key of the French cipher." There is no evidence Keefer ever solved the French cipher. Philip H. Sheridan, *Personal Memoirs of P. H. Sheridan* (New York: Charles C. Webster & Co., 1888), 2:226. Sheridan also sent a copy of Napoleon's dispatch to General Ulysses S. Grant from New Orleans, 12 January 1867, and wrote that the dispatch was genuine. Ulysses S. Grant Papers, Microcopy, Roll 24, LC. Sheridan's copy is in his Papers, R47, LC.

64. Keefer to Seward, New Orleans, 17 January 1867, ibid. Copies of Seward's replies to Keefer have not been located in the Seward Papers nor in the State Department files in the National Archives.

65. Leonard Whitney to George Baker, Washington, D.C., 20 November 1866. RG 59, Records of the Bureau of Account: Miscellaneous Letters Received, Entry 212, NA.

66. Leonard Whitney to George Baker, Washington, D.C., 17 December 1866, ibid. Whitney, cashier for the Western Union Telegraph Company, seemed unconcerned about the huge increase in the monthly bill, for he wrote "Please indicate what corrections, if any, are to be made in bills and return to me and I will send them to you recptd." He would soon learn there was a problem: Baker, the disbursing clerk of the State Department, wrote to him, enclosed money for the December telegraph bill and added "No arrangement has yet been made with the Atlantic Telegraph Co." Cf. Baker to Whitney, 18 January 1867, FIG 59, Records of the Bureau of Account: Miscellaneous Letters Sent, Entry 202, NA. John H. Haswell, "Secret Writing," in *The Century Illustrated Monthly Magazine*, 85 (November

1912), 89. He wrote that the cost of the cable exceeded $23,000. Cf. Fletcher Pratt in *Secret and Urgent: The Story of Codes and Ciphers* (Indianapolis and New York: Bobbs-Merrill, 1939), 191-192, and Clifford Hicks in "Tales from the Black Chambers," *American Heritage*, 24 (April 1973), 58: both authors state $23,000 as the cost. E. Wilder Spaulding, *Ambassadors Ordinary and Extraordinary* (Washington D.C.: Public Affairs, 1961), 72, notes the cost at $13,000; and Bigelow in *Retrospections*, 611, wrote that the State Department was charged somethingover $13,000.

67. Fields received a letter from Baker, which included Seward's request to come to Washington for a discussion of the cable issue: cf. Fields to Seward, New York City, 12 December 1866, Seward Papers, R98, LC. The Hunt Deposition also notes that he and Field went to Washington at Seward's request: Deposition of Wilson G. Hunt, RG 123, B307, NA.

68. Hunter, from Rhode Island, began his service in the State Department in 1829, served under twenty-one different secretaries of state and twelve presidents, and would clerk for more than fifty-five years. Cf. Page proof of Whitelaw Reid's column on Hunter for *The New York Daily Tribune*, which Reid sent Hamilton Fish, 20 May 1879, in the Hamilton Fish Papers, Container 123, LC.

69. Deposition of Cyrus Fields, 23 August 1870, RG 123, B307, NA.

70. Deposition of Wilson G. Hunt, 8 June 1870, ibid. Seward's biographer, Glyndon Van Deusen, found a Machiavellian streak in him, "a love for obfuscating his adversaries by ambiguities that on occasion bewildered even his friends," cf. Seward, 565.

71. Within the next year or so, the State Department owed a total of $32,240.75, and of that amount, Western Union received $933.20; the Anglo-American and other European companies, $20,871.70; and finally, the amount which the New York Newfoundland and London Company should have received was $10.435.85. Cf. Deposition of Henry H. Ward, 5 October 1870. RG 123, B306. NA.

72. Ibid.

73. Hunt and Field to Seward, New York, 27 December 1866 as reprinted in the Petition, RG 59, M179, R319, 34, NA.

74. Seward to Field and Hunt, Washington, 29 December 1866, Seward Papers, R98, LC.

75. The Petition, RG 59, M179, R319, 15, NA.

76. Deposition of Leonard Whitney, 12 October 1870, RG 123, B306, NA.

77. Stoeckl to Gorchakov, Washington, D.C., 25 March 1867, RG 59, Telegrams Sent by the Department of State, 1867-69, Entry 309, National Archives. Hereafter cited as RG 59, E209, NA. Stoeckl would use the State Department telegraph office for two more telegrams (the State Department was charged $49.97) on 22 May and 25 May 1867, when he telegraphed the Russian consul, Martin Klinkowstrvern, in San Francisco and told him the Alaskan Treaty had been ratified by the emperor and thus American ships and merchandise could be landed free in the new northwest American possessions. He also cabled Gorchakov again on 20 June 1867, notifying him ratifications had been exchanged. This time, $69 in gold was paid the same day: cf. RG 59, E209, NA.

78. Stoeckl to Gorchakov, Washington, D.C., 25 March 1867, ibid.

79. Taylor, *Seward*, 278. According to another account, Sumner went to Seward's house, where he learned from Stoeckl and Frederick Seward that a treaty was being prepared: Stoeckl then went to the State Department to meet with Secretary Seward and complete the treaty; however, Sumner went to his own home at 322 I Street. Cf. Van Deusen, *Seward*, 541. Apparently, Seward added the $200,000 to the purchase price on his own authority: cf. Ronald J. Jensen, *The Alaska Purchase and Russian-American Relations* (Seattle and London: University of Washington Press, 1975), 77.

80. Taylor, *Seward*, 277-281; also Van Deusen, *Seward*, 547-[547?] where he states Walker admits receiving $21,000 in gold and $2300 in greenbacks for his services. Jensen, Alaska Purchase, 122-133, describes various interpretations regarding distribution of the funds that were not sent to London for the Russian account.

81. Hunt to Seward, the Petition, FIG 59, M179, R319, NA. Also Seward to Dix, Washington, D.C., 24 May 1867 in RG 59, M77, R58, NA. Seward to Dix and Adams, 24 May 1867 in FIG 59, E209, NA. The cable charges of $111.75 in gold were paid two weeks later, and subsequent cables were also paid within a few days after transmission.

82. Hunt to Seward, New York, 1 June 1867, as reprinted in the Petition, RG 59, M319, R3 19,35-37, NA.

83. Seward to Hunt, Washington, D.C., 11 June 1867, in ibid., 38.

84. Whitney to Baker, Washington, D.C. 19 June 1867, RG 59, E209, NA.

85. Seward to Dix, Washington,D.C., 19 August 1867; same date for dispatch to Clay. RG 59, M77, R58, NA.

86. John Haswell to Hamilton Fish, Washington, D.C., 8 July 1873, Hamilton Fish Papers, R95, LC.

87. State Department Telegrams, 1867-1869, in RG 59, E209, NA reflect the complications posed by this code. Also many other dispatches to and from U. S. ministers during these years contain other examples of this defective code design.

88. Hunt to Seward, New York, 20 November 1867, in the Petition, RG 59, M179, R319, 39. The original letter is in RG59, M179, R267, NA.

89. Hunt to Seward, New York, 20 November 1867 in the Petition, FIG 59, M179, R3 19,40, NA.

90. Seward to Hunt, Washington, D.C., 23 November 1867, in ibid., 41. A copy of Seward's letter is in RG 59, Domestic Letters of the Department of State, M40, R63, NA.

91. Hunt to Seward, New York, 24 January 1868, as reprinted in the Petition, FIG 59, M179, R3 19, 41-42, NA. A copy of the Hunt letter is also found in FIG 59, M179, R271, NA. Enclosed with Hunt's letter was a copy of a letter from George Seward, secretary and general manager of the Atlantic Telegraph Company. Seward wanted Seward to know there were 200 American shareholders in the Atlantic Company and that surely they should not sustain the loss of this revenue. Moreover, the first year's operation of the cable provided only a two percent return on the capital investment: cf. Seward to Hunt, 25 November 1867 in FIG 59, M179, R27, NA.

92. Seward to Hunt, Washington, D.C., 27 January 1868, as reprinted in the Petition, FIG 59, M179, R319, 43, NA. A copy of this letter is in RG 59, M40, R64, NA.

93. Cooper, Taylor, Roberts, Hunt, Field to Seward, New York, 30 January 1868, as reprinted in the Petition, FIG 59, M179, R319, 43-44, NA.

94. Seward to Cooper, Taylor et al., Washington, D.C., 5 February 1868, as reprinted in ibid., 44-46.

95. Hunt to Seward, New York, 12 February 1868 as reprinted in ibid., 46-48. A subsequent letter from the directors to Hamilton Fish on 20 November 1869, recounts the date and mailing of Seward's 24 November letter to Hunt: cf. ibid., 52.

96. Field to Johnson, 10 March 1868, Seward Papers, R 103, LC. Both Johnson and Seward sent telegrams to Field, 10 March 1868, with congratulations and best wishes for the telegraph builders: cf. FIG 59, M40, R 64, NA.

97. Simon Stevens, New York, 5 October 1868, as reprinted in the Petition, FIG 59, M179, R319, 48-49, NA. Hunt's letter, 31 July 1868, and Lathers's reply, 1 August 1868, are also reprinted, ibid., 49-50. A copy of Stevens' letter is in RG 59, M179, R288, NA.

98. Seward to Stevens, 21 November 1868, as reprinted in the Petition, FIG 59, M179, R319, 51, NA. Field continued to maintain a business relationship with Seward, asking him for a few moments to discuss the proposed canal across the Isthmus of Darien and inviting him to a dinner at Delmonico's in honor of Prof. S.F.B. Morse in December 1868, and, finally, sending him a chart with all the principal telegraph lines in operation, under contract, and contemplated for completing the circuit of the globe. Cf. Field to Seward, 7 and 19 December 1868, and 13 August 1870, Seward Papers, FI106, 108,LC.

99. Cooper, Taylor, Field, Roberts, Hunt to Fish, New York, 23 November 1869, in the Petition, RG 59, M179, FI319, 51-53, NA. The Russian cable had been paid 22 August 1868 in Washington, D.C., apparently by the Russian minister after the House of Representatives had voted $7,200,000 for the purchase of Alaska one

month earlier: cf. Memorandum of Account with the Department of State, ibid., 15. Also cf. Deposition of Leonard Whitney, 12 October 1870, RG 123, B306, NA. Page 16 of the Petition stated that the cable to Russia was charged to the Department of State by Mr. Seward's direction. However, Hamilton Fish wrote to the Court of Claims that this statement was not known or believed to be correct, and that the payment for it, if made, was not made by or under the direction of the State Department, nor by any other department or officer of the United States. Cf. Fish to the Court of Claims, Washington, 13 April 1870, RG 59, M40, R67, NA. The "Argument for the Claimant" filed in the U.S. Court of Claims for the December term, 1870, specifies the Russian legation paid the $9,886.50 on 22 August 1868 in Washington, D.C.: cf. FIG 123, B306, NA.

100. Fish to Cooper et al., Washington, D.C., 30 November 1869, as reprinted in the Petition, FIG 59, M179, R3 19, 53-54, NA. Also J. C. B. Davis to the Court of Claims, Washington, D.C., 25 July 1870, FIG 123, B307, NA.

101. The Petition, FIG 59, M179, R3 19,56, NA.

102. "Argument for the Claimant," RG 123, B306, 26, NA.

103. "Brief in Defense" filed in the Court of Claims of the United States, 3 May 1871, RG 123, B306, NA. Also, Seward Deposition, 8 August 1870 in ibid.

104. "Findings of Fact and Conclusions of Law," Court of Claims, 26 May 1871, FIG 123, B307, NA.

105. Ibid.

106. Had payment in gold been stipulated, the cost to the government would have been $35,787 in greenback currency: cf. Mitchell, *Gold*, 316.

107. Taylor to George Boutwell, New York, New York, 6 June 1871, Record Group 217, Accounting Office of the Treasury Department, Office of the First Auditor, Misc. Treasury Account 180406, NA.

108. No. 180406, Comptroller's Office, 28 August 1871, ibid.

Chapter 18

1867 State Department Code

Because of the expensive 23 November 1866, diplomatic cable to John Bigelow in Paris, Secretary William Seward promptly discontinued use of the old Monroe Code and "set to work as early and prosecuted as vigorously as possible the construction of a new and frugal cipher code. . . ."[1] As explained in the State Department introduction to this new code, the magnetic telegraph required the sender to translate code numbers into letters since numerical signs could not be transmitted. Thus, it happened that fifteen to twenty letters were necessary to express a single letter of the old Monroe Code. A determined and chastened Seward wanted a much more economical system for his secret dispatches.

The newly designed code of 1867, based upon the letters of the alphabet and the frequency of the most common words in the English language, often turned into an awkward communications mask for telegraphers and code clerks, as well as diplomats. Designed for frugality, the code required telegraphers to maintain extremely precise spacing between encrypted groups. Although the code appeared efficient and secret in design, it was awkward to use for telegrams and cables, and caused numerous problems for department and legation clerks during the next eight years.

In this code of 148 printed pages, twenty-three words that were the most frequently used in dispatches were assigned one letter of the alphabet. Two other single letters of the alphabet expressed verb tense and plural or genitive third person singular. The letter "w" was not used, except in a cipher table, because it is not used in European languages of a Latin origin and thus would puzzle telegraph operators in those language areas. The next most common 624 words were assigned two letters of the alphabet; three letters for the remainder of the vocabulary required for common diplomatic usage; and a fourth letter was added for plurals and certain parts of verbs. Code symbols were also prepared for the principal countries and cities in the world, for states, major cities, and territories of the United States, and for proper names of men in English. A cipher table was to be used for those words or names not in the code list.

The first seventy-four pages of the code was the encode portion, and it contained the words in alphabetical order together with the code symbols; for example, the very first word was Aaron with its symbol "aba"; the last word on the first page was acknowledge with symbol "ea." To decode a dispatch was a very frustrating and time-consuming task since the three-letter symbols were published in several sequential alphabetical orders. Hence, one had to search through different sections for the plaintext word. This code, designed for economy and cables, did not please telegraphers and code clerks.

More importantly, transmission of the code by cable proved awkward since there was not a standard number of code characters, and sometimes encoded elements were run together by telegraphers. For example, code elements "a" for the and "k" for from might be run together in the cable and appear as "ak," which meant those. American diplomats often transmitted their urgent and secret dispatches by cable. In addition, they also sent them by post, and frequently the State Department could not decode the cable passages until the postal dispatch arrived because of telegraphers' mistakes in spacing the code letter elements.

The transmission problems became so serious that William Seward wrote to the secretary of the Anglo-American Telegraph Company six months after issuing the code and complained about telegrapher mistakes in transmitting encrypted dispatches. He thought there were no problems in the transmission on the route between Washington, D.C., and Heart"s Content, Newfoundland; however, between there and Valentia, Ireland, there had been frequent and important errors. Unaware that the code design invited transmission errors, Seward wrote, "This cannot be ascribed to any complication in the cipher itself, for as that is composed of letters of the alphabet only. . ." but rather was due to telegraphers.[2] And he noted the multiple errors in a recent cable from the State Department to the United States minister at Copenhagen in which code elements were merged. Seward concluded angrily, "Such a result is certainly not calculated to inspire confidence in your medium of communication."

The new code masked communications between the State Department and American legations overseas not only to forestall foreign intelligence agents but also to protect dispatches from domestic interception. Thus, Secretary Hamilton Fish wrote the following dispatch to American diplomat Robert Schenck in London in 1872:[3]

	telegrams	passing	between	you
Thornton complains that the	YNS	OIG	El	BC

and	me	are	published		journals	here	
E	AA	AX	MLHD	in the	CEFS	AG	I think the

leak	is	in	the	telegraph	office		use	the
FYF	I	J	A	YO	QY	You had better	CV	A

cipher	sufficiently		important	despatch	
HMVIL	XIXJ	in each	OR	DXB	at least

obscure	the	meaning
FQG	A	OSF

This thrifty complex code, designed primarily for economy, caused many frustrations for the State Department and the ministers because of mistakes by telegraphers during transmission. Foreign codebreakers must also have been baffled as they sought to decrypt intercepted American dispatches. A State Department clerk and future codemaker, John Haswell, recalled those serious problems in a letter to Hamilton Fish: "It will not perhaps have escaped your recollection, that the first cipher message as received at the Department from our minister to Turkey formed one long string of connected letters, which for a time was considered by many in the Department as a conundrum, but finally, after considerable labor was deciphered by Mr. Davis." In fact, Haswell added, "A telegram was received in a similar condition from Paris, and also one from Vienna. . ." and the latter one was never deciphered.[4] Numerous other such encoded dispatches, beginning in the first months after August 1867, may be found in State Department files.[5] Despite the thoroughly defective design, this State Department code would be used until 1876.

	RUSSIA	NETHERLANDS	GREAT BRITAIN	MEXICO	FRANCE	SPAIN	GERMANY
1866	-	-	11	-	33	-	-
1867	-	-	-	-	-	-	-
1868	-	-	38	-	-	-	-
1869	-	-	122	-	-	26	-
1870	6	-	184	-	27	52	11
1871	259	-	61	-	5	-	40
1872	3	-	189	-	-	-	10
1873	--	-	1	-	1	31	6
1874	17	-	-	-	-	34	2
1875	20	-	-	-	25	20	46
1876	--	-	-	-	13	-	-
Totals	**305**	-	**606**	-	**71**	**170**	**115**

Source: Ralph E. Weber's *United States Diplomatic Codes and Ciphers, 1 775-1938*

Encoded lines in American dispatches from European legations, 1866-1876

KEY

a - c.	j - o.	s - v.
b - f.	k - d.	t - w.
c - h.	1 - b.	u - k.
d - j.	m - g.	v - p.
e - l.	n - y.	w - q.
f - a.	o - r.	x - t.
g - e.	p - u.	y - z.
h - i.	q - n.	z - s.
i - m.	r - x.	

a. The.	aa. Me.	be. Do.
b. It.	ab. Be.	bf. How.
c. Have.	ac. My.	bg. We.
d. {part.,passive,or	ad. At.	bh. Three.
imperfect, indic-	ae. Old.	bi. First.
ative}	af. Now.	bj. By.
e. And.	ag. Here.	bk. This.
f. Of.	ah. So.	bl. Us.
g. Ing.	ai. As.	bm. Far.
h. See - sea.	aj. All.	bn. Second.
i. Is.	ak. Those.	bo. Way.
j. In.	al. Who.	bp. Up.
k. From.	am. Sure.	bq. Plainly
1. But.	an. Will	br. Should.
m. This.	ao. Other.	bs. Yet.
n. That.	ap. Men.	bt. Only
o. To.	aq. Justifiable.	bu. Some
p. On.	ar. Whom.	by. Believe
q. For.	as. No.	bx. Any
r. There.	at. Can.	by. Possible.
s. P1ura1, genitive,	au. Make.	bz. Possibly.
third person sin-	av. Am.	Ca. Learn.
gular, indicative}	ax. Are.	cb. Lead.
t. A.	ay. With.	cc. Country.
u. An.	az. Such.	cd. People.
v. Or.	ba. Would.	cc. Within.
x. If.	bb. Say.	cf. Its.
y. Which.	bc. You.	cg. Offer.
z. Not.	bd. Desire.	ch. Term.

Surprisingly, this defective code would be employed by the State Department until the Cipher of 1876, the Red Cipher, replaced Seward's frugal creation.

Notes

1. Deposition of William Seward, 27 July 1870, Records of the U.S. Court of Claims, General Jurisdiction Case Files, 1855-1937, Case No. 6151, Record Group 123, Box No. 307, National Archives.

2. Seward to John C. Deane, Washington, D.C., 16 January 1868, Record Group 59, Microcopy 40, Roll 63, National Archives. Hereafter cited as RG 59, M40, R63, NA.

3. Fish to Schenck, Washington, D.C., 16 June 1872, Letter Copy Book, 13 March 1871 to 25 November 1872, Fish Papers, Library of Congress. Sir Edward Thornton was the British minister in Washington, D.C.

4. Haswell to Fish, Washington, D.C., 8 July 1873, Microcopy, Roll 95, ibid.

5. For example, Seward to Cassius Clay, Washington, D.C., 30 December 1867, RG 59, M77, R 137, NA. Also, J.C.B. Davis to Fish, Washington, D.C., 16 August 1869, RG 59, E 209, Telegrams sent by the Department of State: many other telegrams in the 1867 code are located in this file.

Chapter 19

Chief Signal Officer's Code for the State Department

A few years after Hamilton Fish became secretary of state in President Ulysses S. Grant's administration in 1869, Colonel Albert J. Myer's office prepared and sent a secret and innovative code to the State Department.[1] The small codebook of eighty-eight pages, measuring seven and three-quarters inches long by four inches wide, contained numerous codewords to mask State Department correspondence. For the very first time, the department had an excellent instrument that provided alternate codes for the hours of the day, the days and dates of the month, the months themselves, and the years. In addition, the book contained over 2,300 codewords spread over seventy-nine pages; however, many of the codewords did not have plaintext words written alongside them.

Another superb innovation in this code was a plan whereby one codeword represented a complete sentence or a lengthy phrase. For example, the codeword "Carbon" would mask the complete sentence "You will charge the necessary expenses on your next account." During the Civil War, Anson Stager had also developed a system whereby arbitrary words represented common expressions such as "I have ordered" and "I think it advisable."[2] This much-improved design had never before been employed by the State Department. Previously, department codes required a word-by-word encryption, and thus codebreakers had a relatively easier assignment.

The codebook contained two columns of printed words on each page plus one center column written in script. Men's names were used mainly for days of the month, river names for the months, animal names for days of the week, cities and countries for numbers, women's names for the hours of the day, and flowers for the years. Listed below are codewords and plain text from several pages of the codebook.

Printed Column	Written in Script	Printed Column
Andrew	7th	Amos
Albert	14th	Henry
Arthur	19th	Frank
Adam	25th	Howe
Allen	1st	George
Abner	11th	Jones
Ben	17th	James
Brown	24th	Lewis
Black	15th	Paul
Bates	31st	Smith
Benton	26th	Hume
Buell	2nd	Hall
Charles	13th	Mason
Calvin	23rd	More
Clark	4th	Grimes
Cameron	9th	Green
Cole	27th	Grant

Chew	3rd	Hunt
David	30th	Ralph
Davis	29th	Norton

Printed Column	**Written in Script**	**Printed Column**
Dawson	5th	Fowler
Day	12th	Martin
Drum	28th	Stanton
Dow	22nd	Thomas
Edward	6th	Ross
Evan	18th	Scott
Emerson	21st	King
Edwin	10th	Knox
Elgin	20th	Sherman
Ewing	16th	Warren
Elias	8th	Newton
Hudson	May	Thames

* * * * *

Mohawk	October	Severn
Santee	June	Gila
Potomac	March	Granges
Rapidan	January	Osage
Platte	April	Oder
Tiber	July	Tagus
Danube	November	Tigris
Nile	December	Humber
Niger	August	Niagara
Rhine	September	Genesee
Seine	February	Red

* * * * *

Cow	Saturday	Rat
Horse	Wednesday	Fox
Goat	Friday	Mule
Lamb	Monday	Lion
Hog	Tuesday	Tiger
Dog	Thursday	Mink
Cat	Sunday	Deer

* * * * *

Rose	1873	Pink
Aster	1871	Peony
Violet	1874	Dahlia
Tulip	1872	Marigold
Daisy		Pansy
Geranium		Sunflower

Printed Column	**Written in Script**	**Printed Column**
France	Sixteen	Leeds
Genoa	Ten	Lima
Georgia	Four	Leghorn
Glasgow	Thirteen	Lepanto
Grenada	One	Madras
Ghent	Eighteen	Madras
Geneva	Two	Mobile
Galveston	Twenty	Memphis
Hayti	One Thousand	Malta
Hamburg	Fifteen	Mecca
Hanover	Seven	Maryland
Havanna (*sic*)	Fourteen	Milan
Halifax	Five Hundred	Minden
Honduras	Three	Montreal
Hungary	Sixty	Moscow
Hull	Thirty	Munster
Indian	Six	Norfolk
India	Ninety	Newark
Italy	Forty	Norway
Ireland	Five	Nashville
Invernes	One Hundred	Nassau
Illinois	Seventeen	Naples
Kent	Eighty	Nantes
Kingston	Twelve	Nubia
Kew	Seventy	Ohio
London	Nineteen	Oporto
Lisbon	Eleven	Paris
Liverpool	Nine	Pekin
Lowell	Fifty	Peru
Lyons	Eight	Palermo

*****

Anna	10:30	Ida
Agnes	12	Jane
Alice	8	Jenny
Amelia	9:30	Kate
Amanda	1	Laura
Betsy	12:30	Lucinda
Bertha	9	Lucy
Clara	3:30	Martha
Catharine	11	Maria
Cornela	1:30	Mary
Clotilda	3	Molly
Delia	11:30	Matilda
Printed Column	**Written in Script**	**Printed Column**
Emily	10	Maggie
Emma	2:30	Nancy

Ellen	5	Nora
Edith	6:30	Nina
Flora	2	Rachael
Fanny	4:30	Rosa
Grace	7	Rebecca
Gertrude	5:30	Susan
Harriet	4	Sarah
Hannah	8:30	Sally
Hilda	7:30	Sarepta
Henrietta	6	Sophia

Printed	**Written**	**Printed**	**Written**
A	D	N	V
B	L	o	R
C	W	P	H
D	C	Q	X
E	I	R	G
F	Q	S	Z
G	o	T	B
H	J	U	F
I	A	V	Y
J	P	W	U
K	N	X	E
L	M	Y	T
M	K	Z	S

Printed Column	**Written in Script**
Achieve	Act of Congress
Acid	Acting Secretary
Acorn	Assistant Secretary
Acrid	Second Assistant Secretary
Across	Admits } singular or
Act	Cannot Admit } plural
Acted	Approval of Congress
Acting	Action
Active	Your action is approved
Acute	Your action is not approved
Actor	Answer by telegraph
Adage	Answer by telegraph in cipher

Printed Column	**Written in Script**
Adapt	Addressed a communication to
Add	Authorize
Adder	You are authorized to
Addle	Appointment
Address	The Senate has confirmed your appointment as

Notes

1. Code Book furnished the State Department by the Chief Signal Officer, U. S. Army, n.d., Hamilton Fish Papers, Container 285, Library of Congress. The document is undated; however, the codebook included flower names for four years, with 1871 as the earliest year noted.

2. William R. Plum, *The Military Telegraph During the Civil War in the United States*, 1:56.

Chapter 20

"Cipher" Dispatches and the Election of 1876

"The story told today by the translation of captured cipher dispatches is not a pleasant one for any American to read," reported a Republican newspaper, the *New York Daily Tribune*, on 8 October 1878. "It is a story of such disgrace and shame that we might well wish that events had not rendered its telling necessary. Every citizen must feel that it would have been better for the good name of the Republic had the contest of 1876, with all its intense passions and its crimes, been permitted to pass from memory." Directed by its famous and especially aggressive editor, Whitelaw Reid, the *Tribune's* journalistic crusade against alleged political corruption by the Democrats sought to uncover the massive electioneering corruption masked in encoded political telegrams. And the daily newspaper would highlight on its front pages all the scandalous maneuvering by leading Democrats during the election of 1876.

Especially targeted were all the encoded messages sent and received by Samuel Tilden's major political advisors and confidants, and maybe even by Tilden himself, at Democratic National Headquarters in New York City. Eagerly judgmental, the *Tribune* proclaimed on 10 September, "It is correspondence in secret cipher – the language familiar to conspirators in crime who dare not face the daylight. Portions translated prove that agents were instructed to buy an electoral vote, and furnished with money to pay for it. Other parts, not yet deciphered, obviously refer to money transactions in immediate connection with the action of returning and canvassing boards and electors at the South."[1]

Samuel Jones Tilden

A few days earlier, the angry Republican editors, reflecting American uneasiness with secrecy, noted that the dispatches were not in the "everyday English of honest men." Moreover, "the very fact that secret ciphers had been arranged before these confidential agents went out indicates that communication was expected of a character which it would not be safe to have known, even to telegraphic operators bound to secrecy."2 The irate editors made no mention of merchants, bankers, foreign diplomats, and journalists, who also used ciphers and codes to protect their confidential messages in the communications world of telegraphs and cables during peacetime.3 Codes and ciphers were required for secure communi-cations because intrigues, collusion, and cabals colored presidential politics at that time. Indeed, politics balanced on the edge between war and peace.

How and why did this newspaper political firestorm about secret messages begin? Who were the major participants?

Was one political party more corrupt than the other? Was the nation, weak, divided, and more susceptible to dishonest politicians because of the chaotic conditions created by the Civil War and Reconstruction?

The presidential election of 7 November 1876, considered the most openly corrupt contest up to that time, found two state governors as candidates: the famous New York reformer of political graft and corruption, Samuel J. Tilden, a Democrat; and Rutherford B. Hayes of Ohio, a Republican, major general of volunteers in the Civil War, former member of Congress, thrice elected as governor – a conscientious leader who pressed for social improvements in the prison system, mental hospitals, and state education. Fifty-three years old, an extremely able administrator, Hayes also gained a modest national reputation as a reformer.

Rutherford B. Hayes

From election returns on 8 November, many persons believed that Tilden had 184 electoral votes, one short of a majority. However, a Republican newspaper, the *New York Daily Tribune*, reported that Tilden had 188 votes, 141 for Hayes, and 34 undecided. When the *Tribune's* pugnacious editor, Whitelaw Reid, learned that the Republican *New York Times* had printed a different set of figures giving Tilden less than a majority of the votes, and reported the race as undecided, a delighted Reid quickly reprinted the *Times's* analysis the following day.[4] Hayes had unquestioned control of 166 electoral votes. In the disputed electoral column were the eight electoral votes of Louisiana, seven of South Carolina, and four of Florida. These were the final three states in which Republican regimes remained strengthened by the votes of blacks. Hayes had carried Oregon; however, the governor of that state, a Democrat, might name a Democrat as a substitute for a Hayes elector because the Republican elector was ineligible to serve since he was a federal officeholder, a postmaster. In the national popular vote, Tilden led his opponent 4,288,546 to 4,034,311.

During the four weeks after 7 November, each political party fought fiercely to ensure victory for its presidential candidate. Because of Reconstruction, a Republican administration, aided by federal troops, dominated the three Southern states and thus hoped to regulate those state Returning Boards that reviewed the election returns for ineligible voters. A majority of the board members were Republican. Could enough Tilden votes be eliminated to award the states to Hayes? Historian C. Vann Woodward argues that both parties employed "irregularities, fraud, intimidation, and violence" during the election.[5] Bitterness and duplicity highlighted this presidential election. As a brilliant team of American historians, Charles and Mary Beard, judged, "By both sides, frauds were probably committed – or at least irregularities so glaring that long afterwards a student of the affair who combined wit with research came to the dispassionate conclusion that the Democrats stole the election in the first place and then the Republicans stole it back."[6]

In the days immediately following the election, representatives of the Democratic

Tilden-Hendricks campaign poster

Hayes-Wheeler campaign poster

Hayes acrobatic poster

and Republican parties went into the three Southern states. These "Visiting Statesmen" maneuvered to bring about or ensure their candidate's victory. And it is during these weeks that the broad flood of cipher telegrams from the Democratic visiting statesmen such as Smith Reed and Manton Marble, former editor and owner of the *New York World*, inundated their New York headquarters. These encrypted documents would later undergo public scrutiny in the congressional investigations.

On 6 December, the Republican electors in the three Southern states met in their capitals and voted for Hayes; Democratic electors also met and cast their votes for Tilden. The Congress was divided on which set of votes to recognize. According to the Constitution, the president of the Senate (then a Republican pro tem) shall open the votes in the presence of the Senate and the House of Representatives; however, the Constitution is silent on whether he or the members of Congress (then predominantly Democratic), acting jointly, should rule on disputed votes. If the Senate president decided on the disputed votes, the Republicans would win; however, if disputed votes were not counted, then neither candidate had a majority, and election would go into the House of Representatives – as it had in 1801 and 1825 – which had a Democratic majority. In addition, if no decision were reached by 4 March, the vice president was to become acting president; however, this office had been vacant since the death of Henry Wilson. The remaining days of December and the first two months of 1877 heightened the dreadful anxieties about the nation's competence to decide on the competing and angry claims to the presidency.

Fortunately for the nation, on 18 January a compromise was achieved and a bill passed that established a fifteen-member electoral commission. The commission would make the final judgment on electoral votes, and it was hoped that this would prevent an armed conflict involving federal troops, national guard units under the control of Democratic governors, and tens of thousands of ex-Union soldiers. As an astute scholar correctly observed, creation of the electoral commission was one of the wisest pieces of statecraft ever evolved by an American Congress.[7] And the real heroes working for the compromise were President Ulysses Grant, Senators George Edmunds, Allen Thurman, Thomas Bayard, Representative George Hoar, and Congressman Abram Hewitt, who also served as chairman of the Democratic National Committee.

Even before the commission deliberated during February on returns for Florida, Louisiana, and South Carolina, Republican agents were in contact with Southern Democrats whose first priority focused on restoration of white power in these states, including withdrawal of federal troops from the Southern states. This objective outweighed the Southern Democrats' quest for the presidency. Quietly, Hayes and his associates assured Southern Democrats that a Republican president would be supportive of their immediate goals.

Beginning on 1 February 1877, each state's electoral vote was tallied in each house of Congress. By an 8-7 vote, the commission awarded Florida to Hayes, and before the month was over, all disputed electors went to Hayes. (In the future, Hayes would be known to his critics as old "Eight to Seven.") Hayes' party leaders were able to prevent a Democratic filibuster in the Congress by promising to remove federal troops from the South, to promote Southern internal improvements, especially a railroad linking the South to the West Coast, and to appoint a Southern leader to his cabinet.

Thus, the electoral vote count was completed at 4 a.m. on 2 March with Hayes receiving 185 votes to Tilden's 184 in the last hours of President Grant's second administration. Hayes took the oath of office privately the evening of 3 March and publicly on 5 March. But many angry Democrats rejected the commission's vote, and on 3 March House Democrats passed a resolution declaring Tilden had been "duly elected president of the United States for the term of four years, commencing on the 4th day of March, A.D. 1877."[8] Clearly, the divided nation and political parties remained on the edge of further turmoil.

Months before the commission voted, the Western Union Telegraph Company had ordered its employees to send to its New York office all dispatches and copies of dispatches relating to the presidential election of 1876. Eager to demonstrate its dedication to maintaining the security of its communications service, the company planned this maneuver to keep the more than 30,000 telegrams, many in cipher, out of the reach of the Congress and publication. They were placed in the care of the company attorney, who would be less likely to be called upon to produce these documents than the other officers of the company.[9] Later, the maneuver failed when the Committee of the House of Representatives on Louisiana Affairs, headed by Democrat W. R. Morrison, called for the Louisiana dispatches. In addition, the Senate Committee on Privileges and Elections requested the Oregon dispatches, numbering 241. The remaining 29,275 dispatches were placed in a trunk and given to the care of the manager of the Washington Western Union office.[10]

The Committee on Privileges and Elections, under the chairmanship of Oliver P. Morton, the Republican senator from Indiana,[11] began taking testimony regarding the electoral votes of certain states in late December 1876 from representatives from New Jersey, Missouri, Minnesota, Oregon, South Carolina, Louisiana, and Florida. The most dedicated committee member was Morton, who, quite naturally, believed the Republicans had won the election. Indeed, after the presidency had been decided, he made a commitment to root out political corruption throughout the entire nation! Later, he traveled west to Oregon to investigate charges of dishonesty against a newly elected senator from Oregon, and while there suffered a stroke in August and died the following November.

As contradictory and highly charged testimony before the Senate committee continued, one aspect of the inquiry came to focus on the Oregon electoral controversy. Right after the election, while the Oregon electoral votes were being disputed, Dr. George L. Miller, editor of the *Omaha Herald*, an aggressive campaigner and member of the National Democratic Committee and close friend of Samuel J. Tilden, was asked by Colonel William T. Pelton of the Democratic National Committee in New York to investigate the Oregon situation in person. Miller, unable to take on this mission, sent a colleague, Mr. J.N.H. Patrick, a lawyer, businessman and zealous Democrat, to see Governor L. F. Grover in Salem, Oregon. Purpose: to acquire one more electoral vote in Oregon.[12]

Before Patrick left Omaha, he and Miller arranged a dictionary code message system. In testimony before the committee, a cautious Miller would not reveal the book's name and noted only that it was small. Moreover, he reported that a memo contained the system for using the dictionary, and without it and the book, he was powerless to explain the system further to the committee. Additional questioning of Miller proved fruitless since Miller, without the dictionary and memo, could not or would not decode the mysterious telegrams shown him by the committee. He testified he had left the dictionary book back in Omaha!

A subpoena by the Senate committee calling for the Oregon dispatches was issued in the latter days of January 1877, and the mysteries of encrypted telegrams, soon to be publicized in the newspapers, would fascinate an anxious and uncertain nation.

These telegrams showed the Democrats employed an exceptional code to veil their hundreds of telegraphic dispatches to Oregon during November and early December. Accounts of the mysterious encoded "Gabble" telegram sent to Tilden on 1 December 1876 from Portland, Oregon, surfaced in the *Detroit Tribune* in early February 1877 and were picked up by the *New York Times* on 8 February.

Incidentally, Republican agents and managers also used a few codes during the exciting days and weeks surrounding the hectic presidential election; however, the few encoded dispatches that surfaced reveal codes of little importance and of simple character. For example, one of the Republican managers, William E. Chandler, a graduate of Harvard Law School and national committeeman from New Hampshire, played a key role in directing strategy during the 1876 presidential campaign. Chandler made only a modest effort to mask a dispatch from Florida:

> **Noyes and Kasson will be here on Monday, and Robinson must go immediately to Philadelphia, and then come here. Can we also have Jones again? Rainy for not more than one tenth of Smith's warm apples. You can imagine what the cold fellows are doing.**

Only five words or phrases were in code: "Robinson" called for $3,000 to be deposited in Philadelphia; "Jones" was $2,000; "Rainy" meant favorable prospects; Smith's "warm apples" actually meant a 250 majority; and "cold fellows" were the opposition Democrats.[13]

Angry Democrats charged that there were few encoded Republican telegrams because William Orton, an avid Republican and the seventy-year-old president of Western Union, permitted party associates to extract some of these telegrams before turning them over to the committee. One scholar complained, "If all the telegrams had been known, it seems probable that Republicans would have been quite as much compromised as Democrats." [14]

The New York Times' editor wrote on 8 February that "Gabble" was a code word for Grover, Oregon's governor, and with humor he explained further that "*The Tribune* has also discovered that where the word 'medicine' is used in Patrick's dispatch it should be translated 'money,' and it may reasonably be inferred that 'Gabble' was a person who could take in a good deal of medicine. In this case it seems to have been prescribed from what the physicians call its "alternative qualities."

In welcome testimony before the elections committee, William Stocking, managing editor of the *Detroit Post* stated that Alfred Shaw came into his office and said he had a translation of the encoded "Gabble" dispatch that he had seen in the Detroit newspapers. With the dictionary, Shaw showed the editor the system for decoding the dispatch.[15] Editor Stocking also paged through the dictionary and confirmed Shaw's findings. Delighted that the Detroit newspaper had publicized the pocket dictionary that supplied the key to the encoded telegrams, the *New York Times* added that the key revealed the following startling information: "The Governor of Oregon informs Tilden five days before he gave his decision that he will decide every point in the case of Post Office Elector in favor of the highest Democratic Elector (E. A. Cronin) and that the certificate will be granted accordingly."[16] The substance of the mysterious Gabble telegram had now been published.

On the evening of 14 February the committee summoned Alfred B. Hinman, an oil merchant from Detroit, who had had

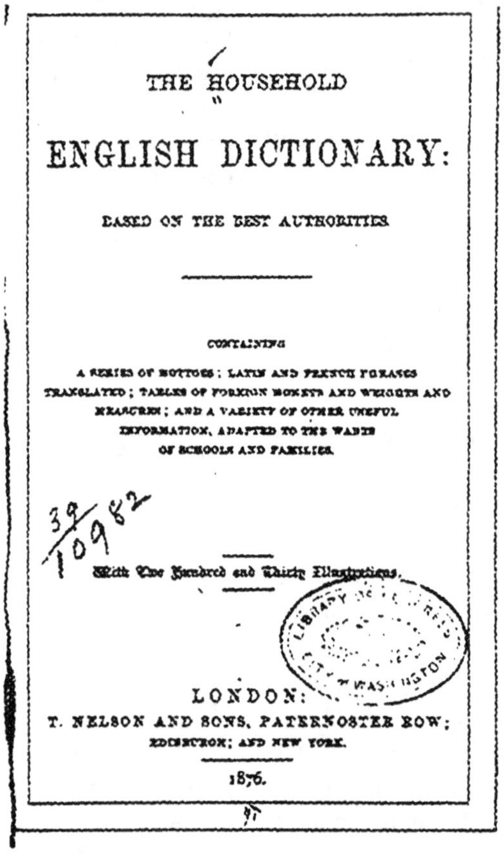
THE HOUSEHOLD

ENGLISH DICTIONARY:

BASED ON THE BEST AUTHORITIES

CONTAINING

A SERIES OF MOTTOES; LATIN AND FRENCH PHRASES TRANSLATED; TABLES OF FOREIGN MONEYS AND WEIGHTS AND MEASURES; AND A VARIETY OF OTHER USEFUL INFORMATION, ADAPTED TO THE WANTS OF SCHOOLS AND FAMILIES.

With One Hundred and Thirty Illustrations.

LONDON:
T. NELSON AND SONS, PATERNOSTER ROW;
EDINBURGH; AND NEW YORK.

1876.

English Dictionary, front page

SPEECH IS THE GIFT OF ALL, BUT THOUGHT OF FEW.

SHE (170) **SHO**

Shac'kle, *v. a.* To chain, to fetter, to link
Shac'kles, *s.* fetters, chains, gyves
Shade, *s.* a shadow; screen, shelter
Shade, *v. a.* to cover from light or heat
Shad'ow, *s. a.* shade, faint representation
Shad'ow, *v. a.* to cloud, darken; represent
Shad'owy, *a.* full of shade; gloomy
Sha'dy, *a.* secure from light or heat; cool
Shaft, *s.* arrow; narrow deep pit; spire

Shed, *v.* to spill, to scatter, to let fall
Sheen, *s.* brightness, splendour – *a.* bright
Sheep, *s.* a well-known animal
Shee'pcot, Shee'pfold, *s.* an enclosure to pen sheep in
Shee'pish, *a.* over-modest, bashful
Shee'pshearing, *s.* the time of shearing sheep; a feast made when a sheep is shorn

Section of *English Dictionary*

previous business dealings with J.N.H. Patrick. He finally revealed that he had exchanged a code system for telegraph messages with Patrick in the summer of 1874 for their private business matters that was based upon the book *The Household English Dictionary*. Based on the best authorities, published in Edinburgh and New York in 1872,[17] this small volume, measuring four inches wide by six inches high, and 3/4 inch thick, contains 241 pages; each page, including those containing sketches of animals or other objects, has two columns.

The most enlightening testimony came from Hinman's general agent and oil merchant, Mr. Alfred W. Shaw, who was most familiar with the system employed by Patrick and Hinman. Moreover, Shaw had already decoded the dispatches that passed back and forth between Democratic committee members in Portland, Oregon, and New York City. Shaw was then asked by the committee to decode the Gabble message below in the presence of the committee:

According to Shaw, who learned the system by trial and error, the key to decoding the Oregon dispatches is as follows: take the word in the dispatch, find it on the proper page and column in the dictionary. Then in the list of words in the column, count the number of words from the top of the page down to the encoded word; and then go forward eight columns (there were two columns per page). In the eighth column, count down the same number of words and you arrive at the plaintext word.[18] Shaw pointed out several exceptions to this rule: the words of and highest do not appear in the dictionary; therefore, they are plaintext words. In addition, the word "accordingly" is also used as it stands because it was customary to use only proper words in these telegrams. Also plaintext words that appear in the first eight columns, such as "a," "act," and "action," are not encoded since they were in the first pages of the dictionary, and therefore the sender could not turn back the necessary eight columns.[19] Shaw said the word "Gabble" means governor, and he learned this from using the different dispatches wherein the word "governor" is the only one that makes sense.

December 1st, 1876

To Hon Samuel J. Tilden, No. 15 Gramercy Park, New York:

Heed scantiness cramp emerge peroration hot-house survivor browse of piamater doltish hot-house exactness of survivor highest cunning doltish afar galvanic survivor by accordingly neglectful merciless of senator incongruent coalesce.

GABBLE

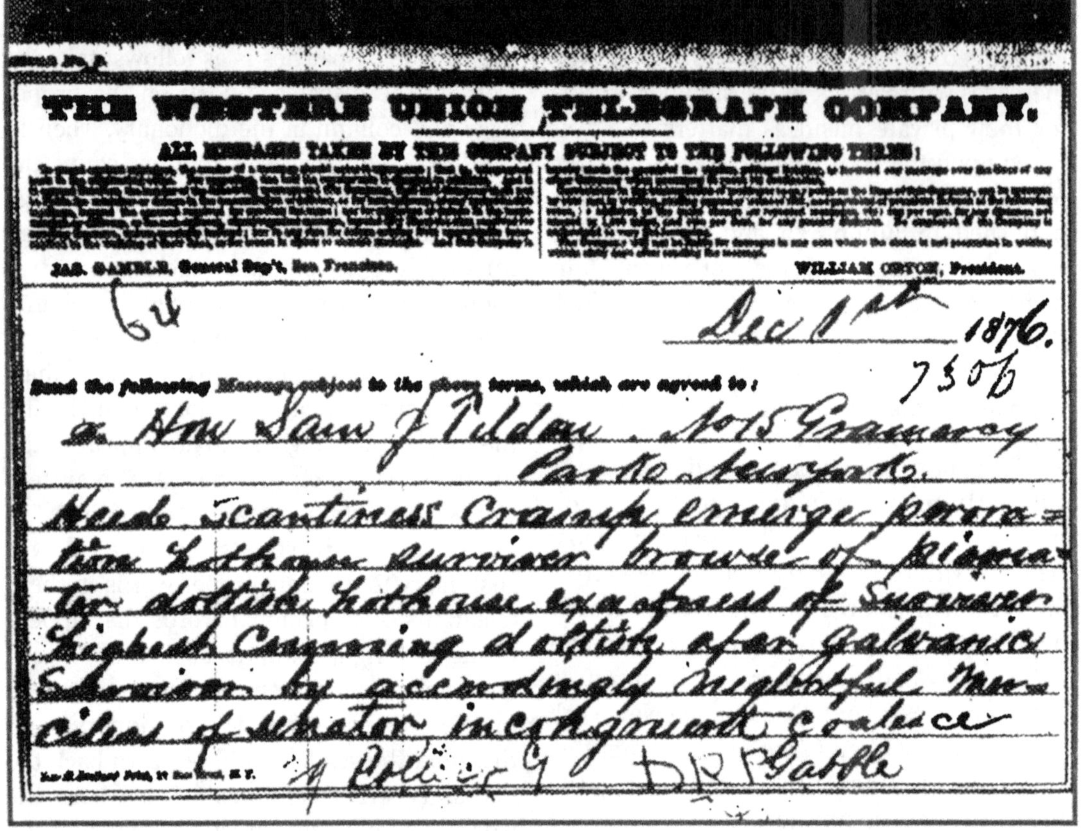

THE WESTERN UNION TELEGRAPH COMPANY.

ALL MESSAGES TAKEN BY THIS COMPANY SUBJECT TO THE FOLLOWING TERMS:

JAS. GAMBLE, General Sup't, San Francisco. WILLIAM ORTON, President.

64

Dec 1st 1876.

Send the following Message subject to the above terms, which are agreed to:

7506

To Hon Sam J Tilden. No 15 Gramercy Park New York.

Heed scantiness cramp emerge peroration hothouse survivor browse of piano tax dottish hothouse exact press of survivor highest cunning dottish afar galvanic sermon by accordingly neglectful merciless of senator incongruous coalesce

DR Gabble

Gabble telegram

Shaw said the encoded Gabble message reads

I shall decide every point in the case of post-office elector in favor of the highest democrat elector, and grant the certificate accordingly on morning of sixth instant. Confidential

Governor[20]

Three of the most interesting encoded telegrams, all of them sent to Tilden's nephew, Colonel Pelton, sent from Oregon and decoded by Shaw, read in plain text as follows:

To W. T. Pelton, Portland, Nov. 28th 1876

No. 15 Gramercy Park, New York:

Certificate will be issue to one democrat; must purchase republican elector to recognize and act with democrat and secure vote and prevent trouble. Deposit ten thousand dollar by credit Kountze Brother, 12 Wall street. Answer.

J. N. H. Patrick[21]

****

To W. T. Pelton, Portland, Oregon, November30

No. 15 Gramercy Park, New York:

Governor all right without reward. Will issue certificate Tuesday. This is a secret. Republicans threaten, if certificate issue, to ignore democrat claim and fill vacancy, thus defeat action of governor. One elector must be paid to recognize democrat, to secure majority. Have employ three, editor only republican paper, as lawyer. Fee, three thousand. Will take five thousand for republican elector. Must raise money; can't make fee contingent. Sail Saturday. Kelly and Bellinger will act. Communicate them. Must act prompt.[22]

****

Portland, December 1, 1876.

W. T. Pelton, No. 15 Gramercy Park, New York: No time to convene legislature. Can manage with four thousand at present. Must have it Monday certain. Have Charles Dimon, one hundred and fifteen Liberty street, telegraph it to Busk, banker, Salem. This will secure democrat vote. All are at work here. Can't fail. Can do no more. Sail morning. Answer Kelly in cipher.[23]

****

Although Shaw decoded sixteen of the telegrams relating to the Oregon dispute word by word in the presence of the committee, certain skeptical senators still questioned the system he used![24]

The Democratic practice of encoding the election dispatches for purposes of secrecy was understandable since their plaintext dispatches often displayed confidential strategies and threatened armed resistance tactics in the bitter political campaign for the presidency. For example, General John M. Corse, an authentic Civil War hero and chairman of the Cook County Democratic Committee in Chicago, sent the following dispatches in plain text (he did not have a code):

To CoL W. T. Pelton, New York, December 6

Glory to God. Hold on to the one electoral vote in Oregon. I have 100,000 men to back it up.[25]

****

To W. T. Pelton, Everett House, New York:

I have no objections to going, but it will take ten days. That will be too long. Can't you send somebody from San Francisco. Just rec'd telegram from Gov. Palmer that vote of Louisiana will be counted for Tilden. Hurrah.[26]

****

Gen. John M. Corse, Palmer House: New York, November 21

If you think it necessary you can pay *National Democrat* [Chicago German newspaper] two hundred and draw on me sight, and thus close it.

W. T. Pelton[27]

Chicago German newspaper

****

Another active Democrat and successful banker in Chicago, W. F. Coolbaugh, sent this distressing telegram in plain text on 14 November to the Democratic leader investigating voting in Louisiana:

To Hon. Lyman Trumbull, St. Charles Hotel, New Orleans:

Should Louisiana republican officials fraudulently change vote, let representative democrats there telegraph governor of Oregon to withhold certificates of election to Hayes electors, and thus protect the people. Telegraph me outlook immediately.[28]

****

November 15, 1876

Perry H. Smith, Saint Charles Hotel, New Orleans, La.

If Louisiana electoral vote is stolen from us, we will get California and Oregon. We have one hundred and sixty thousand ex-soldiers now enrolled. Vast number of republicans with us. Stand firm.

Corse [29]

****

Daniel Cameron, the private secretary of Cyrus H. MCormick, chairman of the State Central Committee of Illinois, joined Corse in sending the following plaintext dispatch to General John M. Palmer, who was also examining Louisiana election returns:

Two hundred thousand ex-Union soldiers, embracing thousands who voted Hayes, sustain you. If Tilden is fraudulently counted out in Louisiana, the end is not yet. You have Illinois behind you.[30]

The chairman of the Democratic National Committee, Abram S. Hewitt, testified to the committee that he was not familiar with *The Household Dictionary*; however, he did explain that in his business he used a cipher dictionary in which he found a five-digit number opposite the plaintext word, and then he added or subtracted from that number in an amount agreed to with his correspondent. Very likely he was using Robert Shaw's Telegraphic Code.[31] Moreover, he added he had not sent a single encoded dispatch during the whole presidential campaign.

W. T. Pelton, Tilden's nephew, who lived at the Tilden residence at 15 Gramercy Park and managed the National Committee office, suffered many amazing lapses of memory before the Senate committee about whether *The Household Dictionary* was the key for the encoded dispatches. He testified that, for the most part, the dispatches were decoded by staff members. He also swore that Governor Tilden never saw the encoded dispatches. When questioned further, Pelton admitted that since the election he had purchased numerous copies of Slater's Code, presumably for business purposes.[32] Pelton replied to dozens of questions about encoded telegrams, either sent or received by saying "I do not remember." Without question, Pelton sat at the nerve center at 15 Gramercy Park and Democratic National Headquarters: it is possible that he or Manton Marble designed or ordered the various complicated code systems for confidential communications between the various Democratic managers in Florida, Louisiana, and South Carolina for their messages back to New York. After Pelton's appearance, the hearings ended on 28 February with over 500 pages of testimony. And, two days later, as noted earlier, the electoral commission determined Hayes had 185 votes.

Tensions about the election persisted for some time as Northern Democrats, including Samuel Tilden, continued to cry "Fraud" when recalling the 1876 election results. Bitter accusations and continuing calls for investigation echoed throughout Congress and the nation.

In the fall of 1878 (two months before the congressional elections), the *New York Daily Tribune* launched a new and at times vicious attack on Governor Tilden, the Democratic party, and their alleged failed attempts to "buy" the election of 1876. The *Tribune* stated that Democratic agents, in seeking electoral votes, offered bribes of $50,000 in Florida, $100,000 in South Carolina, and actually paid $3,000 in Oregon.[33] Packets of encoded Democratic telegrams were sent anonymously to editor Whitelaw Reid during the summer months. These telegrams were part of the 30,000 that had been subpoenaed in early 1877 by the House and Senate committees and later, it was believed, returned to the Western Union Company and destroyed by burning. In fact, however, a large number of the dispatches had been abstracted while in the custody of the Morton committee, and many of them, along with some original telegrams, were eventually given to Reid. The secretary of the National Republican Committee, William E. Chandler, also provided additional copies. The total number exceeded 700 telegrams. Democratic critics would charge that unscrupulous Republicans not only provided the dispatches but also destroyed Republican telegrams that might damage Republican reputations.[34]

The *Tribune* eagerly took up the challenge of investigating these dispatches, which originated in Florida, South Carolina, and Oregon. "The Oregon story is so good an illustration of the insincerity of Democratic professions and the rascality of Democratic practices that it is well worth while to repeat it," wrote the editors. "We have gone through the whole of the vast pile of telegrams relating to this matter – some hundreds in all – and have translated a number of cipher messages that have never before been explained, besides correcting several others which have been imperfectly interpreted." Headlines for the story denounced the "Oregon Fraud: A Full History of the Tilden Plot, How the Democratic Reformer Attempted to Purchase a Majority of the Electoral College – The Cipher Dispatches."[35] This edition and several subsequent ones reprinted copies of the encoded and decoded telegrams on the "Tilden Plot" in Oregon.

One month later, the *Tribune* covered its front pages with another installment of what it termed "The Tilden Ciphers" under a two-column lead, "The Captured Cipher Telegrams." And in an editorial under the heading "The Secret History of 1876," the editors claimed, with some exaggeration, "The most ingenious and intricate system of ciphers yet known was devised in order to conduct the negotiations." The various Democratic cryptographic systems provided concealment far superior to most earlier American cryptographic designs with a major exception: Jefferson's cipher wheel.

The New York Daily Tribune's energetic Whitelaw Reid, then forty years old and an ardent Republican, joined the *Tribune* staff in 1868, and became managing editor the following year. Publishing literary contributions from Mark Twain, Bret Harte, and Richard Henry Stoddard, together with improved foreign news coverage, and comprehensive reporting on the Whiskey Ring scandal, and overthrow of the Canal Ring, increased the circulation and national influence of this daily newspaper. By 1876, the *Tribune* boasted 60,000 readers, mainly conservative, middle and upper class.[36] Though the paper supported Tilden as governor of New York in 1874, it backed Hayes in the 1876 presidential election.

For Reid, the Democratic telegrams provided a marvelous journalistic challenge and a welcome opportunity to weaken the Democratic party. Fortunately, he found a brilliant staff member to investigate the difficult codes. That member would become one of the *Tribune's* successful codebreakers, one of a fascinating trio of brilliant United States codebreakers, probably the most famous cryptographic experts in nineteenth century America: John Hassard, William Grosvenor, and Edward Holden.

John Rose Greene Hassard, then forty-two years old, was a graduate of the Jesuit's St. John's College, Fordham, obtaining both bachelor's and master's degrees. Writing for the *New American Cyclopaedia*, some reporting for the *New York Tribune*, and then preparing a first-rate biography on Archbishop John Hughes were a prelude to his brief editorship of the newly founded *Catholic World* in 1865.

John Rose Greene Hassard, then forty-two years old, also hungered for political reform at both the municipal and national levels. Hassard would argue that telegraphic cipher appeared to be necessary in all important political campaigns. And then he added an appealing but rather impractical idea: "It would hasten the Reform millennium, however, if such messages – being in no right sense of the word private telegrams, but a part of the apparatus of popular elections – could always be collected by Congress after the close of the contest, and exposed to public view, on the ground that the people ought to know exactly how their business has been conducted."[37] He found the Tilden cipher dispatches an attractive and demanding problem. And he plunged into the mysteries of masked messages.[38]

Fortunately, Reid and Hassard found another staff person, the *Tribune's* economic editor, Colonel William Mason Grosvenor, who also delighted in solving challenging problems. The same age as Hassard, Grosvenor had been editor of New Haven's *Journal-Courier*, followed by stints as editor of the *St. Louis Democrat* and manager of reformer Carl Schurz's successful campaign for the United States Senate.[39]

When questioned by Potter's Select Committee on Alleged Frauds in the Presidential Election of 1876, Reid testified that both Hassard and Grosvenor "worked very industriously and zealously" on the enciphered dispatches. "Mr. Hassard, however, did the largest part of the work. He was the earliest in the field and the latest. Both of them did exceedingly good work."[40] And both of them worked independently!

Noting there were no fewer than six distinct systems of cryptography in the secret telegrams, the *Tribune* focused on the scandalous history of the electoral campaign after the election of 1876. Their review of the political correspondence covered almost 400 telegrams, of which about one-half were in plain text and the others in code, between Democratic managers in New York and their secret agents and friends in California, Oregon, Florida, Louisiana, and South Carolina. Their description of the Oregon telegrams described the system for decoding these mysterious messages by employing, as they term it, "The Little Dictionary," which in actuality was *The Household English Dictionary* noted above. It is interesting to note that the *Tribune* did not credit the significant contributions of Alfred Shaw in explaining the code system for the Oregon telegrams.[41]

Another code system, used in the Southern telegrams, provided greater challenges, a pattern of what the editors termed a cipher within a cipher. In reviewing the Southern telegrams, the *Tribune* team noted there were no words for "Democrat" or "Tilden" or "President" or "Hayes" or "Telegram," and they suspected geographical proper names were substituted for these names and other expressions. Their first break came when they focused on the word "Warsaw," which frequently appeared in most of the longer dispatches and in one message appeared all by itself. They suspected it might mean telegraph, and this as confirmed when they worked on the following telegram:

With help from the word "telegram," they determined the words in the dispatch should be ordered according to the following sequence: 9,3,6,1,10,5,2,7,4,8. Their progression was proved correct when they applied the distribution order to many other ten-word telegrams and discovered they could also read them.[42] However, some ten-word telegrams did not translate well with this sequence, and, after trial and error, they discovered a second sequence for other ten-word telegrams: 4,7,2,9,6,3,8,10,1,5. Hassard and Grosvenor provided pages upon pages of decoded dispatches to the *Tribune*.

Another brilliant individual, Professor Edward S. Holden, a mathematician,[43] was chosen by Potter's Select Committee in late January 1879 to examine all the campaign dispatches held by the committee and to decipher them. Holden had become fascinated by the novel and ingenious character of the encoded dispatches in early September 1878, and months before he was solicited by the committee. Apparently he approached the *Tribune* and met with John Hassard, who recalled Reid's desire to seek counsel on the codes from a mathematics professor. Hassard gave Holden several of the dispatches, and later the professor wrote to Hassard about them and requested, perhaps because of his government employment, that his name be kept confidential. According to Reid's testimony, no Holden translations of specific dispatches were received by Hassard or Grosvenor until those dispatches had been decoded by the *Tribune* executives. There

Columbia, Nov. 14, 1876

To Henry Havemeyer, New York:

Warsaw they read all unchanged last are idiots can't situation.

The *Tribune* team translated it as follows:

Can't read last telegram. Situation unchanged. They are all idiots.

The illustration below shows how they found the key:

1	2	3	4	5	6	7	8	9	10
Warsaw	they	read	all	unchanged	last	are	idiots	can't	situation

was one exception: Professor Holden did locate one of the dictionaries that had been used by the Democrats shortly before Reid found it.[44] However, it is interesting to note that Hassard, Grosvenor, and Holden consulted and exchanged opinions with one another about particular codes.

Holden's study and report, prepared between 24 January and 21 February 1879, for the committee, included plain text and decoded copies of the Democratic dispatches. He provided a superb analysis and summary of the different code and cipher systems developed by the Democrats in the election of 1876. This collection of codes and ciphers reflects an amazing quest by energetic Democratic leaders seeking electoral votes: the masks also depict ingenious methods by Democrats for maintaining secret communication. And during the many weeks of testimony before the Senate and House committees, the Democratic leaders such as Marble, Pelton, Weed, and others provided no information on their secret codes and ciphers. Only the congressional investigation on the maneuvers involving the Oregon electoral vote provided information on the dictionary key for those encrypted dispatches.[45]

Holden's thorough report began with a description of the sequence keys used by the Democrats and Republicans for their telegrams.[46] The complicated and highly imaginative cryptographic systems solved by Holden are noted below:

Table of Keys

10	words	15	words	20	words	25	words	30	words
I	II	III	IV	V	VI	VII	VIII	IX	X
9	4	8	3	6	12	6	18	17	4
3	7	4	7	9	18	12	12	30	26
6	2	1	12	3	3	23	6	26	23
1	9	7	3	5	5	18	25	1	15
10	6	13	6	4	4	10	14	11	8
5	3	5	8	13	1	3	1	20	27
2	8	2	4	14	20	17	16	25	16
7	10	6	1	20	16	20	11	5	30
4	1	11	11	19	2	15	21	10	24
8	5	14	15	12	19	19	5	29	9
		9	9	17	13	8	15	27	5
		3	14	1	10	2	2	19	19
		15	5	11	6	24	17	28	17
		12	10	15	7	5	24	24	25
		10	13	18	14	11	9	4	22
				8	17	7	22	7	28
				16	11	13	7	13	1
				2	15	1	4	18	18
				10	9	25	10	12	12
				7	8	22	8	22	6
						9	23	21	21
						16	20	15	20
						21	3	3	29
						14	13	9	14
						4	19	14	7
								2	3
								6	11
								16	13
								23	10
								9	2

There were two different keys for messages of various lengths, and all telegrams were of the segments noted above. In a few instances, the messages were not in a multiple of five words; however, these dispatches probably resulted from mistakes. Moreover, the secret telegrams were transmitted to and received from Democratic representatives in Florida, Louisiana, and South Carolina as they contacted New York, and the particular code these men used often had application only to the specific state, rather than to the covey of states. Also in contrast to the code vocabulary of numbers and names below, the sequence keys are more definitive and subject to proof than the vocabulary plain text drawn from context and deduction.

Code Vocabulary for Numbers

Code	Number	Code	Number
River	0	Potomac	6
Rhine	1	Schuylkill	7
Moselle	2	Mississippi	8
Thames	3	Missouri	9
Hudson	4	Glasgow	hundred
Danube	5	Edinburgh	thousand

Names, Terms

Code	Plaintext	Code	Plaintext
Africa	Chamberlain	Ithaca	Democrats
America	Hampton	Lima	acceptable, ed
Amsterdam	bills[?]	London	canvassing board
Bolivia	proposal	Louis	governor
Brazil	too high[?]	Max	John F. Coyle
Bavaria	probably some Republican official	Monroe	county
Bremen	Commissioner[?]	Paris	draw
Chicago	cost, draft,expense	Petersburg	deposit
Chili [*sic*]	cautious[?]	Portugal	some Republican official: possibly Chandler
Copenhagen	dollars	Rochester	votes
Denmark	Colonel Pelton	Russia	Tilden
Europe	Louisiana	Syracuse	majority
Europe	Gov. Kellogg	Utica	fraud
Fox	C.W. Woolley	Vienna	payable[?]
France	Florida	Warsaw	telegram
France	Gov. Stearns	Asia	?
Greece	Hayes	Dryden	?
Havana	Republicans		

These plaintext words were chosen after reviewing hundreds of telegrams: a majority of the terms were developed by the *Tribune* experts and also found to be in agreement with Holden's translations.

HENRY HAVEMEYER   21 TALLA., FLA 4.
NO. 15 WEST 17TH ST
NEW YORK

1	2	3	4	5	6	7	8	9	10
HALF	TWELVE	MAY	LESS	THIRTY	ELEVEN	WINNING	TEN	ADDITIONAL	SEVEN

11	12	13	14	15
FOR	GIVE	LIEUTENANT	SIXTEEN	RUSSIA

FOX

MAY WINNING [WOOLLEYI GIVE TWELVE (HUNDRED] ELEVEN [THOUSAND] TEN [DOLLARS] LESS HALF FOR RUSSIA [TILDEN] ADDITIONAL SIXTEEN (CANVASSING BOARD] THIRTY-SEVEN [MEMBER] LIEUTENANT NULLI

FOX [C.W. WOOLEY]

key IV: 3, 7, 12, 2, 6, 8, 4, 1, 11, 15, 9, 14, 5, 10, 13.

Copy of Wooley Letter

Dumb Words or "Nulls"

1. Anna
2. Captain
3. Charles
4. Daniel
5. Jane
6. Jones
7. Lieutenant
8. Thomas
9. William

The words above were used to fill out telegrams so the number of words would equal 10, 15, 20, 25, or 30 words so that the proper series noted above could be used.

Number Code

According to Holden, Colonel Grosvenor provided the code noted below. Holden noted the code did not, however, provide a complete plain text for one dispatch.

Code	***Plaintext***	***Code***	***Plaintext***
France	Two	Nineteen	Received
Italy	Three	Twenty	Agree, agreed, agreement
Greece	Four	Twenty-one	Telegraph
England	Five	Twenty-three	Edward Cooper
One	Telegraphic credit	Twenty-four	Vote
Two	Will deposit	Twenty-seven	J.F. Coyle
Three	Supply or provide	Thirty	Republicans
Four	Have you arranged or deposited	Thirty-two	Canvassing
Five	Will send, or remit	Thirty-four	G.P. Raney
Seven	Draft or draw	Thirty-five	Requirements
Nine	Bank	Thirty-seven	Member
Ten Dollars	Forty	Expenses	
Eleven	Thousand	Forty-one	Paid or protected, accepted
Twelve	Hundred	Forty-six	Prompt, or prudent
Sixteen	Canvassing Board	Fifty	

John Hassard, noted Holden, worked out part of the South Carolina code below:

Code	***Plaintext***
Bath	Court
Cuba	Electoral vote of South Carolina
Naples	Majority
January	Democratic
Jo	Telegraph
April	Failure
Chicago	Cost, expense

The testimony of William E. Chandler revealed that the Republicans- code was indeed modest and also somewhat creative as they termed the Democrats "Cold Fellows!" Holden noted the following list:

Code	***Plaintext***
William	Send
Rainy	Things look favorable
Robinson	$3,000
Jones	$2,000
Brown	$1,000?
Smith	$250?
Warm apples	Majority
Cold Fellows	Democrats
Oranges	Florida
Cotton	Louisiana
S.C.cotton	South Carolina

Dictionary Codes

Holden included another dictionary in addition to *The English Household Dictionary.* He found that *Webster's Pocket Dictionary* contained the word "Geodesy," which proved to be the key for decoding four dispatches.

Double-Number Code

20	d	62	x
25	k	66	a
27	s	68	f
31	1	75	b
33	n	77	g
34	w	82	i
39	p	84	c
42	r	87	v
44	h	89	y
48	t	93	e
52	u	96	m
55	o	99	j

Cipher

Cipher	n o p q r s t a b c d e f u v w x y z g h i j k l m
English	a b c d e f g h i j k i m n o p q r s t u v w x y z

Other Ciphers

English	Key A	Key B	Key C	Key D	Key E	Key F	Double Letter
a	n	-	p	1	h	z	yy
b	o	-	t	-	i	a	ma
c	p	-	e	o	j	b	ep
d	q	-	r	n	k	c	it
e	r	z?	-	c	1	d	ns
f	s?	-	o	p	m	e	ye
g	t?	-	n	k	t	f	mm
h	a	i?	k	y	u	g	pp
i	b	-	I	v	v	h	ei
j	c	-	-	-	w	i	-
k	d	-	q	-	x	j	ia
e	m?	a	-	y	k	sh	
m	f	-	b	-	z	1	ny
n	u	-	d	g	a	m	ss
o	v	w?	c	f	b	n	aa
p	w	-	f	a	c	o	sn
q	x?	-	w	-	d	p	-
r	y	i?	u	d	e	q	pi
s	z		-	e	f	r	im
t	g	-	v	b	g	s	pe
u	h	c?	z	r	n	t	ai
v	i	-	i	t	o	u	em
w	j	-	y	-	p	v	sp
x	k?	-	s	s	q	w	-
y	1	-	h	u	r	x	en
z	m	-	-	-	S	y	-

Key B and Key D above are so fragmentary because only one message was sent in each cipher. Only three messages were sent in Key F; and the Double Letter Key was deduced from the translations given in the *New York Daily Tribune*.

Many years later, the brilliant cryptographer William F. Friedman reconstructed the square upon which the double letter and double number ciphers were based. And the key phrase used by the Democratic agents is most interesting in the light of the charges of bribery! With this square, Friedman also supplied the missing combination for the letter J, which is nn, and X,which is yi. Holden's report did not include those two letters. Q and Z had no ciphers in either solution by Holden or Friedman. The square developed by Friedman is as follows:[47]

		H	I	S	P	A	Y	M	E	N	T
		1	2	3	4	5	6	7	8	9	0
H	1										
I	2					K		S			D
S	3	L		N	W					P	
P	4		R		H				T		
A	5		U			O					
Y	6		X				A			F	
M	7					B		G			
E	8		I		C			V		Y	
N	9		E			M				J	
T	0										

Samuel Tilden requested an appearance before the cipher subcommittee of Potter's Select Committee when it was holding its sessions in New York City's Fifth Avenue Hotel in February 1879. Earlier, his associates, Colonel Pelton, Manton Marble, and Smith Weed, had testified about the Democratic involvement with the electoral votes. On the 8th, in a circus-like atmosphere, unruly curiosity seekers, together with officials, politicians, and representatives of the press, clogged the hallway leading to the committee room two hours before Tilden's scheduled appearance. At 11:30, a weak but defiant Tilden appeared, dressed in black, with "an air of great solemnity on his face, which looked as imperturbable and sphinx-like as ever," the *New York Herald* reported.[48] "Since his last public appearance, he seemed to have aged considerably, and yesterday he looked quite ill and feeble. As he afterward explained, he was suffering from a severe cold. It was, indeed, quite a painful spectacle to see the slow, halting, lame walk with which he passed the table and reached his seat. His figure was stiffly drawn up and seemed incapable of bending, as though he were suffering from a paralytic contraction of the limbs. . . . Not a muscle of his face relaxed with animation or expression as he stiffly extended his hand. . ." to the two Republican members of the committee, and after saluting the three Democratic members, he "took off his elegant, silk-lined overcoat, stiffly turned round and seated himself at the table, while settling at the same time a large handkerchief in his breast pocket."

His sober, highly rational testimony, lasting over two and one-half hours, emphasized emphatically that he had no knowledge, information or suspicion of cipher correspondence until it appeared in the *Tribune*: "I had no cipher; I could not read a cipher; I could not translate into a cipher." When saying this, he hit the table with his clenched fist. When he referred to the bribes alluded to in the cipher dispatches, his faint hoarse voice became loud, vehement, and dramatic: his face flushed and "the mental excitement had such mastery over him that his lips twitched, and one of his hands, said to be smitten with paralysis, trembled in a most painful manner."[49] He had not selected or sent the Democratic "Visiting Statesmen" to the South. Firmly he asserted, "No offer, no negotiation, in behalf of any member of the Returning Board of South Carolina, of the Board of State Canvassers of Florida, or of any other State was ever entertained by me, or by my authority or with my sanction. No negotiation with them, no dealing with them, no dealing with any one of them was ever authorized or sanctioned by me in any manner whatsoever."[49]

With righteous anger and frustration, he declared, "to the people who, as I believe, elected me President of the United States; to the four million and a quarter of citizens who gave me their suffrages, I owed duty, service, and every honorable sacrifice, but not a surrender of one jot or tittle of my sense of right or of personal self-respect" exclaimed an embittered Tilden. "I was resolved that if there was to be an auction of the Chief Magistracy of my country, I would not be among the bidders."[50] Goaded by Republican congressman Frank Hiscock's bitter and intensive questioning, Tilden finally exploded: "I declare before God and my country that it is my entire belief that the votes and certificates of Florida and Louisiana were bought, and that the Presidency was controlled by their purchase." And replying to Hiscock's request for proof, Tilden argued that the committee's investigation had sufficient evidence for this declaration. After a few more hostile questions from Hiscock and G. B. Reed, Republican of Maine, Tilden was excused: he had had his final day in court.

Tilden's testimony closely reflected a popular Democratic view: that the electoral votes of Florida, Louisiana, and South Carolina were for sale, that a few of Tilden's closest friends knew this and at a minimum were not averse to negotiating a purchase, but they did not buy them. Rather, Republicans secured the votes. In addition, Republican managers got access to the telegraphic dispatches of both parties, destroyed their own, and publicized those of the Democrats. From the time the Hayes administration took over in March 1877, Returning Board members and all other Republicans associated with the election returns in the disputed states had been rewarded with offices. And finally, according to this Democratic perspective, the Republicans, with "virtuous indignation," held up the dispatches to prove that Tilden was a "fellow who wanted to steal but was not smart enough."[51]

Tilden's testimony, together with Edward Holden's report on the cipher dispatches, completed the investigative phase of the Potter Committee's inquiry on the alleged electoral frauds in the 1876 presidential election. Over 200 witnesses had been examined, over 3,000 pages of testimony published. The committee's majority report on 3 March 1879 reflected the views of the seven Democrats: (1) the Canvassing Board of Florida reversed and annulled the choice of its residents; (2) the choice of the people of Louisiana was annulled and reversed by the Returning Board; and (3) Samuel J. Tilden, not Rutherford B. Hayes, was the real choice of a majority of the electors duly appointed by the several states and of the voters.[52]

The minority report argued that no evidence was presented as to the dishonesty of the canvassing boards in Florida, Louisiana, and South Carolina. Rather, the *Tribune* publication of the cipher dispatches showed that the very men who had been loudest in their denunciations of the boards had tried to corrupt the electoral process in those states with money. According to the evidence, the minority report specified that no Republican dispatches were intentionally destroyed, and that the innocence of Democratic leaders such as Colonel Pelton and Samuel Tilden was not established.[53] These reports concluded the painful series of electoral investigations begun over two years earlier: the two major political parties remained bitterly divided. The next presidential nominating conventions were fifteen months ahead. Although Tilden believed he could be nominated and elected in 1880, he wrote on 18 June to the New York delegates at the National Democratic Convention in Cincinnati: "In renouncing my renomination . . . it is a renunciation of reelection. . . . To those who think my . . . reelection indispensable to an effective vindication of the right of the people to elect their own rulers. . I have accorded all along a reserve of my decision as possible, but I cannot overcome my repugnance to enter a new engagement which involves four years of ceaseless toil . . . such a work is now, I fear, beyond my strength."[54]

Was Tilden correct in writing of "reelection" to the office of the president? Had he actually won the popular balloting in 1876 only to lose the electoral vote? He deeply believed this. And much historical evidence supports this interpretation. Had there been a fair election in South Carolina and Louisiana in which Black voters were protected, Hayes would have won those states. In Florida, under similar conditions, Tilden would have triumphed.[55]

The cipher telegram congressional investigations and the newspaper publicity generated by Whitelaw Reid in the *New York Daily Tribune* had clearly exhausted the leading Democratic candidate. Indeed, a proud Reid, soon after the story broke in 1878, predicted the cipher revelations "have made an effective end of any political future he may have had."[56] The election strains left Tilden with a shattered body, a slow shuffling gait, and increased "numb palsy" or paralysis agitans: in sum, an old broken man at the age of 66.[57] And perhaps the final irony to the story of the cipher telegrams and the election of 1876 is to be found in the successful Republican candidate in the election of 1880. The presidential victor, Ohio congressman James A. Garfield, had been a member of the 1877 electoral commission!

Notes

1. *New York Daily Tribune*, 10 September 1878.

2. Ibid., 5 September 1878.

3. The editor of the *New York Herald* on 10 October 1878 printed a more balanced interpretation regarding secret communications: "The fact that they were in cipher affords no reasonable presumption against the honor and integrity of either receivers or senders. Politicians on both sides had an interest in concealing their movements from their adversaries. The resort to a cipher was no offence against political morality so long as the objects attempted to be accomplished were legitimate."

4. Bingham Duncan, *Whitelaw Reid: Journalist, Politician, Diplomat* (Athens: University of Georgia Press, 1975), 68.

5. C. Vann Woodward, Reunion and Reaction: *The Compromise of 1877 and The End of Reconstruction* (New York: Doubleday & Co., 1951), 19.

6. Charles A. Beard & Mary R. Beard, *The Rise of American Civilization* (New York: MacMillan Co., 1942), 2:3 14.

7. Paul Leland Haworth, *The Hayes-Tilden Disputed Presidential Election of 1876* (Cleveland: The Burrows Brothers Company, 1906), 333. Historian Bernard A. Weisberger made the wise observation that the compromise was "engineered by the businessmen of both parties, quintessential conservatives, who did not want the already struggling economy hit with political paralysis or renewed warfare." Cf. Bernard A. Weisberger, *American Heritage* 41 (July-August 1990): 19.

8. As quoted in Hodding Carter, *The Angry Scar: The Story of Reconstruction* (New York: Doubleday & Co., 1959), 342.

9. U.S., Congress, House Select Committee on Alleged Frauds in the Late Presidential Election, Cipher Telegrams, 45th Congress, 3d session, 1879, H. Rept. 140, part 2, 1 (hereafter cited as Cipher Telegrams, H. Rept 140).

10. Ibid., 22.

11. U.S., Congress, Senate Sub-Committee of the Committee on Privileges and Elections, Testimony on the Electoral Vote of Certain States, 44th Congress, 2d session, 1877, Misc. Doe. 44 (Washington, D.C.: Government Printing Office, 1877), 1 (hereafter cited as Testimony, Misc. Doe. 44).

12. Ibid., 425-26.

13. John R.S. Hassard, "Cryptography in Politics," *North American Review* 128 (March 1879), 319.

14. Alexander Clarence Flick, Samuel Jones *Tilden: A Study in Political Sagacity* (Port Washington, New York: Kennikat Press, 1963), 429-30.

15. Testimony, Misc. Doe. 44,440.

16. *New York Times*, 8 February 1877. *The Times* identified the Detroit newspaper as the *Tribune*.

17. Ibid., 439. This particular edition is not located in the Library of Congress nor in the British Museum catalog that lists a similar volume, though published in London and New York, 1872. However, the Library of Congress does have *The Household English Dictionary*, published in Edinburgh and New York, for 1876 but not 1872. This edition may have been the book used by Patrick, Miller and Pelton for their masked dispatches since it provides the key for decoding the messages.

18. Ibid., 442. A special Detroit dispatch to the *New York Times*, 8 February 1877, incorrectly described the dictionary as having three columns on each page and also mistakenly said that the sender turned back two pages to choose the corresponding word, counting from the top of the page. *The New York Daily Tribune*, 7 October 1878, made the same error of specifying four pages instead of eight columns. The four-page description is repeated in Hassard, "Cryptography," *North American Review* 128:322.

19. Testimony, Misc. Doe. No.44,449-451.

20. Ibid., 448.

21. Ibid., 448.

22. Ibid., 449.

23. Ibid., 456.

24. Ibid., 442-58.

25. Ibid., 407. In testifying, Corse doubted he sent the telegram; however, if he did, he stated, it was "mere badinage." Cf. Testimony, Misc. Doc. 44,407.

26. Ibid., 409.

27. Ibid., 409.

28. Ibid., 410.

29. Ibid., 411.

30. Ibid., 412.

31. Ibid., 490-91.

32. Ibid., 504-05.

33. *New York Daily Tribune*, 22 June 1880.

34. Cipher Telegrams, H. Rept. 140, 2. Also cf. U.S. Congress, House Select Committee on Alleged Frauds, 45th Congress, 3d session, Misc. Doe. 31 Part IV (Washington D.C.: Government Printing Office, 1879) 4:110 (hereafter cited as House Misc. Doe. 31).

35. *New York Daily Tribune*, 4 September 1878.

36. Duncan, Whitelaw Reid, 248.

37. Hassard, "Cryptography," *North American Review*, 128:315-25.

38. Allen Johnson and Dumas Malone, eds., *Dictionary of American Biography* (New York: Charles Scribner's Sons, 1931), 6:382-383.

39. Ibid., 6:26.

40. House Misc. Doc.31,4:111.

41. Whitelaw Reid in testimony before the House Committee in January 1879 stated "I think that the key to the Oregon dispatches was first discovered by somebody in the West, and that it had been published in some Western paper." Ibid., 4:111.

42. Ibid., 7 October 1878.

43. Dumas Malone, ed., *Dictionary of American Biography* (New York: Charles Seribner's Sons, 1932), 5:136-37.

44. House Misc. Doe. 31,4:112.

45. Testimony on the Electoral Vote of Certain States contains lengthy testimony from William Pelton, George L. Miller, John M. Corse, and three dozen others involved in the Oregon case during hearings between December 1876 and March 1877: cf, Testimony, Misc. Doc. 44. Much more extensive testimony, provided in January and February 1879, involving manipulations in Oregon and the Southern states is to be found in the *House Select Committee on Alleged Frauds* documents: cf. House Misc. Doc. 31, Part 4.

46. The Holden report on his reconstruction of the codes and ciphers may be found in the House Misc. Doc. 31, 4:325-33 1.

47. William F. Friedman and Lambros D. Callimahos, *Military Cryptanalytics* (Washington, D.C.: National Security Agency, 1956), 1:97-99.

48. *New York Herald*, 9 February 1879.

49. House Misc. Doc. 3 1,4:273.

50. Ibid., 4:274.

51. *New York Herald*, 10 February 1879. John Bigelow, editor, author, and diplomat, and Tilden's close friend and biographer, was much more harsh in his indictment: ". . . during the whole four years of Hayes' administration, and

regardless of Mr. Tilden's age, his physical infirmities, his priceless public service, and the place which he occupied in the hearts of his countrymen, he was pursued by the agents of that administration with a cruelty, a vindictiveness, an insensibility to all the promptings of Christian charity, which men are rarely accustomed to exhibit except in their dealing with wild beasts." John Bigelow, *The Life of Samuel J. Tilden* (New York: Harper & Brother, 1895), 2:169.

52. House Report 140,67.

53. Ibid., 69-77.

54. Flick, *Tilden*, 456.

55. Flick, *Tilden*, 415. H. J. Eckenrode, *Rutherford B. Hayes: Statesman of Reunion* (New York: Dodd, Mead & Co., 1930), concludes that if Florida votes had been honestly tabulated, Tilden won the presidency, and the electoral votes should have been Tilden 188 to Hayes 181: 193. Vann Woodward in *Reunion and Reaction*, 20, agrees that recent historical scholarship holds that Hayes was probably entitled to the electoral votes of Louisiana and South Carolina; however, Tilden should have been given Florida's four electoral votes, thus giving him 188 votes and the presidency. Pulitzer-prize-winning Allan Nevins, also Leon B. Richardson and William Dunning, awards Florida to Tilden. A notable dissenter, Paul Maworth, in his *Hayes-Tilden* book, argues that in a free election, the three Southern states, Florida, Louisiana, and South Carolina, would have returned substantial majorities for Hayes and therefore the election of Hayes was the proper one, 330-32,341.

56. As quoted in Duncan, *Whitelaw Reid*, 71.

57. Flick, *Tilden*, gives a good medical analysis of Tilden's health, 417.

Chapter 21

John H. Haswell: Codemaker

"It will not perhaps have escaped your recollection that the first cipher message as received at the Department from our minister to Turkey formed one long string of connected letters, which for a time was considered by many in the Department as a conundrum."

In the 1870s John Haswell renewed the pioneering cryptographic endeavors of Charles W. F. Dumas, James Lovell, and Edmund Randolph, who inaugurated unique systems for masking U.S. diplomatic correspondence one hundred years earlier. Sensitive to the innovative world of cable messages, Haswell recognized the necessity for providing American post-Civil War diplomats with an efficient, secure, and economical communications instrument. He was born in Albany, New York, on 7 February 1841, to Henry Burhans Haswell and Elizabeth Trowbridge, the third of seven children.[1] His grandfather, John Haswell, Sr., had left Northumberland, England, in 1774 and settled on a farm near Bethlehem, six miles southwest of Albany. The future State Department chief of the Bureau of Indexes and Archives studied at the Albany Boys' Academy, became a member of the Albany Zouaves Cadets, and later graduated from the recently established Law School of Georgetown University in 1873.[2]

The young John Haswell was appointed to a temporary clerkship in the State Department on 23 January 1865, by another New York resident, Secretary of State William Seward. Promotions followed rapidly: Haswell became a Clerk Class One in August 1867 (salary, $1,200); Class Two in March 1869; Class Three, June 1870; and Class Four, June 1871 (salary, $1,800). His clerical assignments included processing departmental diplomatic correspondence and preparing special reports for the Congress.[3] After the assassination of President Abraham Lincoln, Haswell carefully gathered the voluminous correspondence, official resolutions, and tributes from writers all over the world, and edited these writings for *The Tributes of the Nations to Abraham Lincoln*.[4]

Secretary Seward assigned Haswell the major responsibility for studying all the various memoranda relating to the purchase of Alaska and for writing the final treaty document for that spectacular acquisition. Also during those hectic months, he processed all the correspondence relating to the Alabama claims against Great Britain.[5] The United States government sought $15 million from the British for the depredations wreaked on Northern shipping during the Civil War by warships, including the Alabama, constructed in Great Britain for the Confederate government.

In processing American diplomatic correspondence, Haswell faced the frustrating problem of decoding secret dispatches that were prepared in the design-defective and economy-driven 1867 code. Domestic and overseas telegraphers frequently changed the spacing between the one-, two-, or three-element code symbols; as a result, State Department clerks were forced to spend hours finding solutions for these puzzling dispatches. Frequently, a secret dispatch could be understood only several weeks later when the department received a copy of the plaintext dispatch that had been mailed from the originating American embassy in London, Paris, Madrid, or another of the European capitals.

The astute and efficient administrator, Secretary of State Hamilton Fish, on 7 August 1873, rewarded John Haswells industry, efficiency, and management skills by appointing him chief of the Bureau of Indexes and Archives in the Department of State with a salary of $2,400.[6] Fish had begun reforming the department's office procedures in June 1870 by creating a small central files unit that processed the essential correspondence from the Diplomatic, Consular, and Home Bureaus. In 1873, the central files unit became the Bureau of Indexes and Archives, managed by three clerks and a chief. "The reorganization recognized for the first time the importance of record keeping and placed the record desks under the direction of an experienced and competent chief, John Haswell."[7] The bureau became the depository for the archives, dispatches and other correspondence to the department (except letters relating to passports and applications for office) and for the other official records. In addition, it was responsible for opening the mail, abstracting and indexing all department correspondence. The job descriptions did not specify the processing and production of codes and ciphers.[8]

The basic divisions of the department's correspondence continued unchanged; however, changes were initiated in registering and recording correspondence. Separate registers were used for each class of letters sent and received, and six classes were established: Instructions to U. S. Ministers and Notes to Foreign Legations; Dispatches from U.S. Ministers and Notes from Foreign Legations; Instructions to Consuls; Dispatches from Consuls; Miscellaneous Letters Received; and Miscellaneous Letters Sent. Before leaving the department in 1894, Haswell also designed an indexing plan in which there was a subject index along with the former arrangement of organizing all correspondence according to source or destination. The most laborious task of the bureau, making handwritten copies of dispatches for the files, continued until 1898 even though a Fairbanks Morse typewriter was added in 1880. Carbon copies became a procedure in 1909.[9]

Frustrated with the telegrapher mistakes when transmitting dispatches encrypted with the 1867 code, Haswell became determined to develop a more efficient and economical code for secret communications. As he wrote, "Formerly, ciphers were used only in the transmission of secret communications and secrecy was the only element considered. Since the introduction of telegraphy, owing to the frequency of communication, expense, another motive, has been introduced and must be taken into account."[10] Thoughtfully, he explained his under-standing of economical communications security: "In a cipher two elements should be associated, secrecy and economy. Secrecy so that our minister might know what the true position of the government as to any particular subject that he might be enabled to make the proper concessions, and other reasons that will readily suggest themselves. Economy so as to enable the sender of a message to fully express his view, without contraction...."[11]

In July 1873, just a month before being named bureau chief, Haswell wrote to Hamilton Fish and described a complex codebook he had developed. He recalled an encoded message, encrypted in the 1867 code, from the American minister in Turkey, and how the dispatch formed one long string of connected letters. The assistant secretary of state, John Chandler Bancroft Davis, after many hours, perhaps days, of careful decryption, finally discovered the plain text. Haswell also explained that cables with similar mistakes arrived from Paris and Vienna, with the latter probably never decrypted.[12]

Because of these frustrating difficulties with code letters and groups being merged, Haswell determined the best solution was to use the same number of digits-five-for each plaintext word or phrase. Then, if a telegrapher neglected the proper spacing, the recipient could simply divide the string of digits into groups of five to obtain the code number.[13] If the sender used only consonant letters, then only four should be used to represent a plaintext word or phrase. He believed the communications convention at Vienna had determined five figures or five letters would equal one word for tariff purposes, and this design had become common to all countries.

Ever mindful of the department's emphasis on economy, Haswell pointed out that the 1867 code had one great advantage over his design: with the earlier code, the clerk could encode twenty-three words by using but one letter for each word; with the use of two letters, express 624 words; and with three letters, communicate the remainder of the vocabulary. An important economical feature of his new design, however, permitted the sender to use only four letters to encrypt the name, for example, of Sir Edward Thornton, the British minister in the United States. The 1867 code required the spelling out of that name with a transposition cipher, at a greater cost.

Haswell emphasized one of the most economical and security-conscious elements in the new design: as suggested in General Albert Myer's plan, which had been sent to Fish, arbitrary words could be used to represent words or phrases and for a route cipher. Anson Stager had developed the idea of using an arbitrary word to represent a phrase of several words such as "I have ordered," and "I think it advisable" during the Civil War.[14] In addition to the Stager and Myer systems, Haswell proudly stressed, his proposed clever design included numerous words that represent general sentences, and one word could mask several words.[15] Thus, the Haswell plan had four distinct and essentially different codes in one book: words, letters, figures, and route.

But the most fascinating and significant new element in Haswell's code was an aperiodic changing key. As he proudly wrote,

> **By a secret understanding between the Department and our Legations, the following would form another [code]. A card could be furnished with the names, say of animals etc. each one of which would change the cipher entirely. i.e. Suppose the word 'cat' should indicate on the card 'read the tenth line ahead,' then by sending 'cat' as the first word in the message, the figure code 189,54 would be read by the person receiving the message as 189,64. The designations on the card could be changed at the pleasure of the Department, or an understanding could be had that as the months changed, the names on the cards would change.**[16]

For the first time in the history of State Department cryptology, an imaginative blueprint offered an encryption system that could be modified by time frame and automatically.

An energetic and thorough manager, Haswell studied the needs of the State Department scientifically: "In the preparation of the present cipher [The 1876 Cipher of the Department of State] both elements [secrecy and economy] were considered. In order to prove its economic character an actual count was made of the plain words used in the cablegrams sent to our Minister at Madrid during the first month." With pride, he added, "The following was the result: number of words in messages, 1630; number of words actually sent, 1024, and the cost was $388.66, for a savings of over 36 per cent."

Fascinated by archival work, and finding State Department records very deficient, Haswell consulted congressional records, also "the great arbiter, the ancient books of the Treasury Department" and spent ten years preparing a complete list of those who directly or indirectly were involved in the foreign relations of the United States."[17]

The four-volume manuscript was entitled "Chronological History of the Department of State and the Foreign Relations of the Government from September 5, 1774, to the present time." This history, which included the names of all the consular officers of the United States, was prepared outside his regular duties for use by the heads of departments, officers in the foreign service, and committees of Congress.[18] Secretary of State Frederick T. Frelinghuysen recommended to the Congress that this "bird's-eye view of men and events" be purchased for $6,000. Congress acted favorably on 7 July 1884, and Secretary Frelinghuysen wrote in early 1885 to Hugh McCulloch, secretary of the treasury, and requested McCulloch to recommend to the Congress an appropriation for the purchase. He added that Haswell had extended the study to 1 July 1885. Despite the endorsements and accolades for his pioneering manuscript, the appropriation was not secured. Secretaries of State James G. Blaine and Elihu Root would add their endorsements for the manuscript's purchase, but to no avail.

Beginning in 1906, Haswell's heirs petitioned the State Department, urging that the government purchase the history manuscript.[19] Several years later, another law firm for the Haswell estate, Penfield and Penfield, in Washington, D.C., promoted another strategy. These lawyers urged that an item for appropriating the funds be inserted in the Diplomatic and Consular Bill, and that the likelihood would be high that when the bill was reported out, no objection will be raised to this item. Thus it would pass the House in this form (the House had previously objected to a bill containing a specific appropriation for the manuscript), and no problem was expected in the Senate since twice previously this body had approved the purchase.[20] However, this tactic also failed, and as late as 1916 the Washington law firm offered little prospect for purchase of the manuscript by the government. Congressional opponents continued to argue that Haswell prepared the study on government time and therefore no payment should be made.[21]

The four volumes remained in the Stationery Room of the State Department until 1927 when Tyler Dennett suggested to the assistant secretary of state, Wilbur J. Carr, that the volumes be re-bound.[22] Carr promptly replied that the manuscript belonged to the estate of John H. Haswell, that it was offered to the government but the appropriation was never made. Thus, Carr recommended the volumes be turned over to the estate's lawyer, T. John Newton, in Washington, D.C., in order to avoid a claim by the estate for compensation for the use of the volumes, or possible damages for their loss or destruction.[23] On 17 May 1927, T. John Newton accepted and signed for three volumes entitled "Consular Officers of the United States, 1775-1893" and one volume, "Consular Officers of the United States, October 21, 1781-October 21, 1881."[24]

Despite his discouragement over the delays in the purchase of his history manuscript, Haswell continued his duties as chief of the Bureau of Indexes and Archives. In January 1893, Secretary of State John W. Foster appointed Haswell as special messenger to the state of Montana and instructed him to obtain promptly the certificate of electoral votes for president and vice president of the United States and to return to Washington, D.C., by the most speedy route, keeping the department advised by telegraph of his movements.[25] The following year, Haswell received a Diploma of

Honorable Mention for his assistance in producing a noteworthy exhibit in the Columbian Exposition.[26] Haswell would resign his post as bureau chief in the State Department in June 1894, during President Grover Cleveland's second administration, when Richard Olney was secretary of state. During his thirty years' service in the department, Haswell served only five years under Democratic administrations.

Sometime during these bureau years, Haswell probably wrote the interesting article "Secret Writing: The Ciphers of the Ancients, and Some of Those in Modern Use" in which he surveyed the art of various forms of confidential writing from ancient times down to the late nineteenth century. This is one of the first articles on secret writing written by an American codemaker. Cryptography, cryptology, polygraphy, stenography, and ciphers are defined in this significant article as he traces the varieties and compositions of secret written communication down through the Middle Ages, into the American republic during its founding, and on to the Civil War. The essay also touches on post-Civil War cryptographic systems of the Navy, War, and State Departments. Moreover, it mentions his personal involvement in the famous William Seward cablegram to John Bigelow, the American minister in France. The "Secret Writing" treatise is a pivotal American contribution to a public understanding of cryptology.[27]

Haswell's final contribution to cryptology was made in 1899 when he delivered a new code, the Blue Cipher, to the State Department. He had convinced Secretary John Sherman that the 1876 codebook had probably been compromised. Once again, his urgent warnings about the need for masking State Department communications reflected his sensitivity to espionage activities supported by foreign governments. Though he had never served overseas, he realized that foreign intelligence services read other people's cables. Before Haswell's death on 14 November 1899, his two codebooks, the 1876 Red Code and the 1899 Blue Code, established State Department encryption systems for secret cable communications, systems that would continue to evolve through World War I and the postwar period, concluding with the Brown code, published in 1938.

Notes

1. Samuel Burhans, Jr., *Burhans Genealogy: Descendants from the First Ancestor in America, Jacob Burhans, 1660, and His Son, Jan Burhans, 1663 to 1893* (New York: Printed for Private Distribution, 1894), 230.

2. W.J. Maxwell, ed., *General Register of Georgetown University* (Washington, D.C.: n.p., 1916), 175. The Law School opened in 1870.

3. John Haswell, "Officers Connected with the Department of State and the Diplomatic Service of the Government, 25 October 1774 to 31 December 1882," 179. This excellent reference manuscript of 825 handwritten pages lists all the officers associated with the diplomatic service of the United States. The obituary in *The Times Union* (Albany, New York), 15 November 1899, incorrectly noted he became a clerk in 1863.

4. *The Times Union* (Albany, New York), 15 November 1899.

5. Ibid.

6. Photostat copy of appointment letter in author's possession. The original letter is in the possession of Mrs. Lester Thayer in Albany, New York. Mrs. Thayer is a granddaughter of John H. Haswell's oldest sister, Margaret Mathilda. Secretary Fish continued to administer an expanding department carefully. The department relocated from the Orphan Asylum Building (where it had been since 1867) to Mullett's Building near the White House in 1875; the office staff increased from twenty-five to forty-five clerks. As historian Allan Nevins wrote, "No Secretary ever had more confidence and loyalty from his associates." Cf. Allan Nevins, *Hamilton Fish: The Inner History of the Grant Administration* (New York: Frederick Ungar

Publishing Co., 1936), 2:864.

7. H. Stephen Helton, "Recordkeeping in the Department of State, 1789-1956," National Archives Accessions, No. 56 (1961), 5. Also cf. John A. De Novo, "The Enigmatic Alvey A. Adee and American Foreign Relations, 1870-1924," *Prologue* (Summer 1975), 7:69-80. Haswell, "Officers," 69.

8. Haswell, "Officers," 68. Soon after becoming chief of the bureau, Haswell reported on the following monthly correspondence categories and volume of the office: diplomatic, 463; consular, 442; pardons and commissions, 86; domestic records, 340; passports, 362; library, 55 volumes issued and 59 returned; and translations, 97 pages.

9. Ibid., 6-7.

10. John H. Haswell to John Sherman, Washington, D.C., 26 January 1898. Photostat in possession of the author; the original letter is owned by Mrs. Lester Thayer of Albany, New York.

11. Ibid.

12. Haswell to Fish, Washington, D.C., 8 July 1873, Container 95, Hamilton Fish Papers, Library of Congress.

13. He also wrote that four letters could form a code group; however, the 1876 code he finally devised used codewords of varying lengths. Cf. ibid.

14. Plum, *The Military Telegraph During the Civil War in the United States*, 1:56.

15. At this time, merchants were also eager to have one code word represent several words or a sentence. Indeed, one mercantile code word, "unholy," represented a record 160 words. Cf. Charles Bright, *Submarine Telegraphy: Their History, Construction, and Working* (London: Crosby Lockwood & Son, 1898), 176.

16. Haswell to Fish, Washington, D.C., 8 July 1873, Container 95, Fish Papers, Library of Congress.

17. Haswell to Fish, Washington, D.C., 29 September 1883, Container 141, Fish Papers, Library of Congress.

18. U.S. Congress, House, "Letter from The Secretary of State . . . recommending purchase of 'Chronological History'," House Ex. Doc. No. 124, 48th Cong., 1st sess., 1884, 1-4.

19. Chandler P. Anderson to Elihu Root, New York, 26 November 1906. Copy in author's possession.

20. Penfield and Penfield to Mrs. Gibson Oliver, Washington, D.C., 10 February 1910. Copy in author's possession.

21. Penfield and Penfield to Mrs. Laura Townsend Oliver, Washington, D.C., 4 February 1916. Copy in author's possession.

22. Dennett to Carr, Washington, D.C., 12 May 1927, Record Group 59, 811.412/219, Box 7530, National Archives.

23. Carr to Dennett, 14 May 1927, RG 59, 811.412/219, NA. Newton Memorandum, 17 May 1927, RG 59, 811.412/219, NA.

24. Cf. also Carr to Newton, Washington, D.C., 14 May 1927, ibid. Sometime after that date, these valuable volumes were returned to the government and, though not accessioned; they are shelved in the State Department section of the National Archives.

25. John W. Foster to John H. Haswell, Washington, D.C., 23 January 1893. Copy in author's possession.

26. Virginia C. Meredith to John H. Haswell, Chicago, 30 November 1894. Copy in author's possession.

27. John H. Haswell, "Secret Writing," *The Century Illustrated Monthly Magazine*, 85 (November 1912), 83-92.

Chapter 22

The Red Code of the Department of State, 1876

In 1876, the year of America's Centennial, an epoch-making secret codebook, *The Cipher of the Department of State*, designed by John H. Haswell and printed in the Government Printing Office, became the initial volume in a series of elaborate State Department codebooks. It established the basic blueprint for secret American diplomatic correspondence for more than six decades. Much of Haswell's research for this codebook, soon termed the Red Code of 1876, was completed before Secretary of State Hamilton Fish appointed him chief of the Bureau of Indexes and Archives in August 1873.[1]

Undoubtedly, Haswell's earlier accomplishments as a department clerk for Secretary William Seward, together with Seward's costly cablegram, made him particularly sensitive to the dire need for an economical secret codebook. Not only did this complex codebook furnish code words and code numbers, but it also added various message transmission routes for further protecting diplomatic dispatches from foreign government post offices and espionage agents. In the State Department, the meticulously prepared volume quickly replaced the 1867 codebook and all the earlier codesheets and ciphers.[2] Colonial codemaker James Lovell would have been delighted to have such an instrument for protecting American diplomatic correspondence.

The innovative and imaginative one-part codebook listed plaintext words, phrases, and short sentences in alphabetical order. The code words were also in alphabetical order, and the code numbers followed alongside in sequential order. This arrangement, though avoiding the need for two distinct books, i.e., one for encode and another for decode purposes, was a distinct security weakness. Almost 1,200 pages in length, this volume contained several approaches for encoding: first, the use of arbitrary words to express plaintext words and sentences, and, second, a system of code numbers that could be used instead of the code words. Thus, the masked message could be sent in code words or five-digit code numbers similar to the design in Robert Slater's codebook, and the plain text, or as the codebook termed it, the "true reading," would be carefully masked.

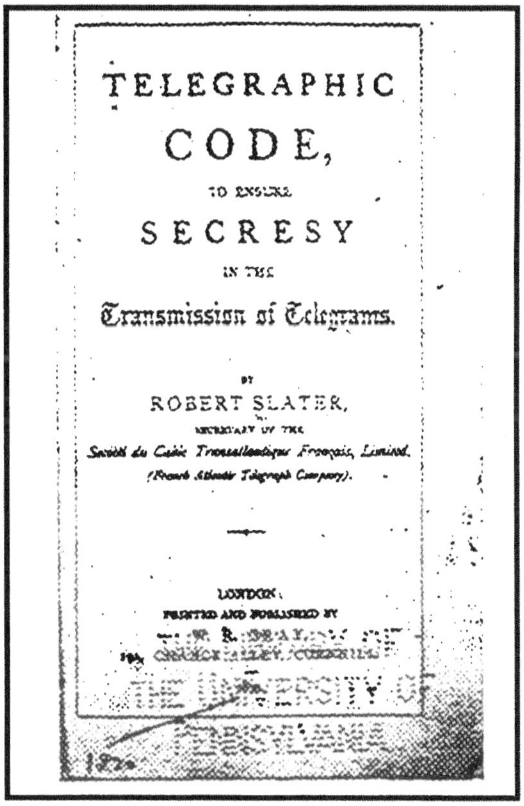
TELEGRAPHIC
CODE,
SECRESY
Transmission of Telegrams.
ROBERT SLATER,

Slater codebook, front page

Numerous State Department frustrations with the 1867 codebook were reflected in the new codebook's "Directions" page, where it was lamented that Chinese, French, German, Italian, Japanese, Russian, Spanish, and

Turkish telegraph operators, ignorant of English, "constantly commit vexatious and often serious mutilations of original messages." According to Haswell, these operators could, however, transmit code-numbers with accuracy because the figures were readily intelligible.

John Haswell and a cost-conscious State Department, still scarred by the exorbitant $19,540 cablegram sent by William Seward to John Bigelow in France in 1866, focused on economy as a prime feature in this complex codebook. In the directions for encoding a message, Haswell carefully explained: "It would be well to commence with the first word therein, and though there be found no expression under it which could be applied to the message, nevertheless find its corresponding codeword and write it down, and proceed in like manner with the remaining words in the message, until there is found an expression which contains some of the words that have already been coded. In this case expunge from the message thus far coded all the code-words found in that expression, and substitute for them the code-word of the expression last found. Thus in sending by means of one word an expression which contains several words, economy, which is one of the principal features of the code, will be secured."[3]

During the exciting decades after the invention of the telegraph, entrepreneurs, merchants, bankers, and governments turned increasingly to codes and ciphers for a modicum of security and lower expenses. The first private secure code for telegrams, along with an explanatory book of reference, was

-1-

1 A	(1)	ABR	
A	00001	Abides	00051
Aback	00002	Abiding	00052
Abacus	00003	Ability	00053
Abaft	00004	Abject	00054
Abalienate	00005	Abjection	00055
Abandon	00006	Abjectly	00056
Abandoned	00007	Abjectness	00057
Abandonce	00008	Abjudicate	00058
Abandoner	00009	Abjuration	00059
Abandonment	00010	Abjure	00060
Abase	00011	Abjurer	00061
Abasement	00012	Ablation	00062
Abash	00013	Ablative	00063
Abate	00014	Ablaze	00064
Abated	00015	Able	00065
Abates	00016	Abluent	00066
Abattoir	00017	Ablution	00067
Abbacy	00018	Ably	00068
Abbe	00019	Abnegate	00069
Abbess	00020	Abnegation	00070
Abbey	00021	Abnormal	00071
Abbot	00022	Aboard	00072
Abbreviate	00023	Abode	00073
Abbreviated	00024	Abodement	00074
Abbreviation	00025	Abolish	00075
Abbreviature	00026	Abolished	00076
Abdicate	00027	Abolisher	00077
Abdicated	00028	Abolishes	00078
Abdicates	00029	Abolishing	00079
Abdicating	00030	Abolition	00080
Abdication	00031	Abolitionist	00081
Abdomen	00032	Abominable	00082
Abdominal	00033	Abominably	00083
Abduce	00034	Abominate	00084
Abduct	00035	Abominated	00085
Abduction	00036	Abominates	00086
Aberrance	00037	Abomination	00087
Aberrant	00038	Aboriginal	00088
Abet	00039	Aborigines	00089
Abetment	00040	Abortion	00090
Abeyance	00041	Abortive	00091
Abgregate	00042	Abound	00092
Abhor	00043	Abounding	00093
Abhorred	00044	About	00094
Abhorrence	00045	Above	00095
Abhorrently	00046	Abrade	00096
Abhorrer	00047	Abrahamic	00097
Abhorring	00048	Abrasion	00098
Abhors	00049	Abreast	00099
Abide	00050	Abrenunciation	00100

-1-

ABR	(1)	ACC	2
Abridge	00101	Abstracter	00151
Abridged	00102	Abstraction	00152
Abridger	00103	Abstractive	00153
Abridgement	00104	Abstractly	00154
Abroad	00105	Abstruse	00155
Abrogate	00106	Abstrusely	00156
Abrogated	00107	Absurd	00157
Abrupt	00108	Absurdity	00158
Abruptly	00109	Absurdly	00159
Abscess	00110	Abundance	00160
Abscond	00111	Abundant	00161
Absconded	00112	Abundantly	00162
Absconding	00113	Abuse	00163
Absconds	00114	Abused	00164
Absence	00115	Abuser	00165
Absent	00116	Abuses	00166
Absentee	00117	Abusive	00167
Absenter	00118	Abusing	00168
Absist	00119	Abut	00169
Absisted	00120	Abutment	00170
Absolute	00121	Abuts	00171
Absolutely	00122	Abutting	00172
Absolution	00123	Abyss	00173
Absolutism	00124	Acacia	00174
Absolutory	00125	Academic	00175
Absolve	00126	Academical	00176
Absolved	00127	Academician	00177
Absolver	00128	Academist	00178
Absolving	00129	Academy	00179
Absonant	00130	Acataleptic	00180
Absorb	00131	Accede	00181
Absorbable	00132	Acceded	00182
Absorbed	00133	Accedes	00183
Absorbent	00134	Acceding	00184
Absorbing	00135	Accelerate	00185
Absorbs	00136	Accelerated	00186
Absorption	00137	Accelerates	00187
Abstain	00138	Accelerating	00188
Abstained	00139	Acceleration	00189
Abstaining	00140	Accelerative	00190
Abstains	00141	Accent	00191
Abstemious	00142	Accentual	00192
Abstention	00143	Accentuate	00193
Absterge	00144	Accentuation	00194
Abstergent	00145	Accept	00195
Abstinence	00146	Acceptable	00196
Abstorted	00147	Acceptance	00197
Abstract	00148	Accepted	00198
Abstracted	00149	Accepter	00199
Abstractedly	00150	Accepting	00200

Slater book, second page

probably devised by the founder of Reuter's Telegraph Company, Baron Paul Julius von Reuter who, after a brief stint as bank clerk, started a pigeon post service in 1849 that filled the span between telegraph stations in Aachen, Germany, and Verviers, Belgium. Soon after, he moved to England, became a naturalized citizen and opened Reuters, a news office in London, which provided financial data for bankers. Within two decades, he added news reports to financial coverage and cabled this intelligence to European and American newspapers eager for current information. In subsequent generations, Reuters became a principal company in an enormous information industry.[4]

Sir Francis Bolton introduced another codebook in 1866. In the formative years, cables proved to be the most expensive branch of the telegraph business. For example, in July 1866 the New York, Newfoundland and London Telegraph Company charged $100 gold for twenty words or less, including address, date and signature, for cables from America to England. Every additional word, not exceeding five letters, cost $5. Code and cipher messages were charged at a double rate.[5] After 1 November 1866, Atlantic cable rates were reduced 50 percent.

Because of the expensive charges, and also the need for secrecy, telegraphers developed new codes.[6] At the International Telegraph Conference in Rome in 1870, and again at the St. Petersburg Conference in 1875, it was specified that code words must not contain more than ten characters (a character is one letter); and words of length greater than ten were liable to be refused, although some companies accepted them and promptly charged higher rates. The Bureau of the St. Petersburg Conference received authority to compile a vocabulary of words to be recognized and accepted for code purposes. However, transmission of encoded messages could be suppressed by the government that granted the cable concession. Indeed, President Woodrow Wilson invoked this measure in the early years of World War I and permitted only the sending and receipt of encoded messages on figures, telegraphers in each repeating station through which the message was sent were urged to repeat back in order to achieve greater accuracy. This meant double service for the telegraph company, and double charges.

In a rapidly expanding domestic market, code words proved especially attractive to retail merchants and other businessmen in the decades after the Civil War. In the early years, various mercantile houses and news organizations built up amazing code vocabularies. In time, almost all commercial traffic was encoded, and some highly imaginative designs were developed. For example, the code word "unholy" was used to designate 160 words. Other examples from this flourishing shorthand code vocabulary were as shown below.

ELGIN	Every article is of good quality that we have shipped to you.
STANDISH	Unable to obtain any advances on bills of lading
PENISTONE	Cannot make an offer; name lowest price you can sell at
COALVILLE	Give immediate attention to my letter
GRANTHAM	What time shall we get the Queen's Speech?
GLOUCESTER	Parliamentary news this evening of importance
FORFAR	At the moment of going to press we received the following[7]

In 1870, Robert Slater, secretary of the French Atlantic Telegraph Company, developed and published one of the most extensive and notable codebooks, entitled *Telegraphic Code, to Ensure Secresy* [*sic*] *in the Transmission of Telegrams*, printed by W. R. Gray in London. Slater noted that the telegraph system throughout the United Kingdom would pass into the hands of the government on 1 February 1870, and Post Office officials would work the lines. "In other words, those who have hitherto so judiciously and satisfactorily managed the delivery of our sealed letters will in future be entrusted also with the transmission and delivery of our open letters in the shape of telegraphic communications, which will thus be exposed not only to the gaze of public officials, but from the necessity of the case must be read by them." Troubled by the greater threat to community privacy, Slater added, "Now in large or small communities (particularly perhaps in the latter) there are always to be found prying spirits, curious as to the affairs of their neighbours, which they think they can manage so much better than the parties chiefly interested, and proverbially inclined to gossip." [8] In addition, he wrote, experience had shown that in the transmission of commercial intelligence it was necessary "to conceal the news communicated from all but the receivers of the messages, and particularly is this the case in the instance of submarine cables. . . ."[9] Finally, he concluded, codes can result in much lower costs.

Secrecy and economy characterized his codebook, which offered 24,000 words (words beginning with C totaled 2,800) in its vocabulary in addition to 1,000 more words expressing Christian names, common surnames, heroes, deities, and some geographical names. Each word was expressed by a five-digit number. This codebook also offered an additional plan for adding or subtracting specific numbers from the code number in order to mask the message further between two correspondents who agreed on the additive number.

In the United States State Department, Hamilton Fish continued to employ the economical 1867 code for foreign correspondence despite the frustrating errors caused by domestic and foreign telegraphers. Searching for better secret writing systems, Fish was influenced by Colonel Albert Myer, a brilliant New Yorker, born in 1829, who entered the army in 1854 as a medical officer and served in Texas. Knowledgeable about the telegraphic code, he transferred the sound system to visual signaling. During the Civil War, Myer organized the Signal Corps and also supervised the construction of 5,000 miles of telegraph lines to western frontier forts.[10] Myer suggested a route or word transposition system, including arbitrary words, to Secretary Fish.

Colonel A. J. Myer

The route system originated with young Anson Stager, another New York state resident, four years younger than Myer, who would become intrigued with the expanding telegraph business. He began his working life as a printer's devil in an office under Henry O'Reilly (who became a leader in telegraph construction and management) and then bookkeeper for a small newspaper, before becoming a telegraph operator in Philadelphia, then Lancaster, Pennsylvania. Rapid promotions followed: after a brief time as telegraph office manager in Pittsburgh, he became, in his early thirties, general superintendent of the Western Union Telegraph Company, with headquarters in Cleveland, Ohio. Stager's early employment made him sympathetic with newsmen and their relations with the telegraph companies. Also, he convinced railroad executives that their companies could profit handsomely by permitting his company to share use of the railroad telegraph lines.

Soon after the outbreak of the Civil War, Captain Stager took over responsibility for all the telegraph lines in the Ohio military district and prepared a cipher for the governor to communicate with the chief executives in Indiana and Illinois. At General George McClellan's Cincinnati home, Stager developed a similar cipher that the general and also detective Allen C. Pinkerton would use. Stager accompanied McClellan's forces and established the first system of field telegraphs used in the war: "The wire followed the army headquarters wherever that went, and the enemy were confounded by the constant and instant communications kept up between the Union army in the field and the Union government at home."[11] When McClellan traveled to Washington to take command of the Army of the Potomac, Stager was assigned to organize the military telegraph in that region; and in early 1862, when the president took control of all the telegraph lines in the United States, Stager became chief of the United States Military Telegraph. "The cryptography used throughout the war was perfected by him, and baffled all attempts of the enemy to translate it."[12]

Stager's encryption system called for determining the number of lines in the proposed telegram and indicating the number by the first word in the message. For example, "mail" indicated one line; "may," two lines; "August," three lines; and so on with descriptors for up to thirty-three lines. His clever design also included check or meaningless words to be included as every sixth word in the message, for example, "charge," "change," "scamp," "thief," and "puppy." The third element in the procedure listed code words for specific officers and locations, such as "Mecca" for McClellan; "Arabia" for Grant, and "Joe" for Martinsburg. The route and number of columns for encrypting and decrypting the message were communicated verbally.[13] The Stager design functioned very well during the Civil War years. And Stager prospered on his return to civilian life: he became superintendent of the Central Region of the Western Union Company, with headquarters in Cleveland, and after 1868, in Chicago.

As early as 1871, and probably because of Myer's persuasion, Hamilton Fish, in his third year as secretary of state, adopted the Stager system for some of his domestic correspondence. He described the system to General Horace Porter, a West Point graduate and Congressional Medal of Honor holder, who was serving as President Ulysses S. Grant's military secretary. Fish wrote: "The plan of the route cypher of which we spoke is simple [-] on receiving a despatch you write off the words in four columns omitting every sixth word then read up the first down the fourth up the third down the second." [14]

Fish continued:

To encypher first count the number of words - divide by 4 and make as many lines as the divided will yield and also one line for any fraction beyond precise division. For instance, of 29 words, make 8 lines then beginning at the bottom of the first column write upward 8 words, then down on the fourth then up on the third and down on the second, filling in any blank spaces on the bottom of the second column with

blind words. Then insert in every sixth place a word to be rejected in decyphering.[15]

continued to use this system over the next several years.[18]

Fish's column design for the route plan appears as follows:

is	down	first	simple
spoke (twice)	the	the	on
we	fourth (Greeley)	up	receiving
which	up	read (British)	a
of	the	then	despatch
cypher	third	word	you
route (going)	down	sixth	write
the	the (Senate)	every	off
of	second	omitting (ground)	the
plan	wooden	columns	words (spoken)
the (indirect)	tribunal	four	in

The encrypted dispatch would be transmitted as follows:

Is down first simple spoke twice the the on we fourth Greeley up receiving which up read British a of the then despatch cypher third word you route going down sixth write the the Senate every off of second omitting ground the plan wooden columns words spoken the indirect tribunal four in

While vacationing at his home in Garrison, New York, Fish began using this route system in 1871 for correspondence with his assistants, John Chandler Bancroft Davis and R.S. Chews, in the State Department. Davis also used the plan for masking important State Department information in his telegrams to Fish. In a letter that probably included his first use of the cipher, Fish also wrote, "What a grand cipher we have!"[16] And several days later, Chews wrote back, "Yes our cipher is a grand one, if its only mission is to make a man crazy."[17] However, despite early difficulties, including telegrapher mistakes, the correspondents

****

As Haswell began fashioning the new Cipher of the Department of State, he studied the first generation of telegraph codebooks and Civil War systems. His carefully designed book would incorporate some of the very best secret techniques in the telegraph and cable industry.

The example printed in Haswell's codebook directions revealed the very detailed, cost-conscious design inherent in this thick volume. To encode the phrase "The President directs me to enter a protest against the rules proposed by the International Congress," the sender referred to the first principal word "President"; and in the codebook a plaintext phrase, "the President directs me" was found for which the codeword was "Plant." Since this code word was on page 443, line 84, the codenumber was "44384." For the plaintext word *protest*, a phrase containing this word in its proper context was found in the codebook as *enter a protest* and was designated by "Precursors" and the codenumber, "45202." For the word *against*, the expression *against the rules* was masked by "Applauded" and "11769": for *proposed*, the phrase *proposed by the* was designated by "Pater" and "45082." Finally, under the word *congress*, the expression *international congress* was found and designated by "Declamation" and "21920." Thus the complete message in codewords read "Plant, Precursors, Applauded, Prater, Declamation." Using code numbers, the message would be "44384, 45202, 11769, 45082, 21920." Haswell emphasized with enthusiasm and genuine pride that a plaintext dispatch of sixteen words could be encoded in five words or five groups of numbers. This codebook provided for economical masked dispatches. Haswell made no mention of communications security.

Instructions for the codebook noted that when the sender could not locate phrases or sentences in the codebook, the plaintext message might be altered so that codebook phrases could be used. If, however, the proposed alteration changed the sense of the dispatch, the sender should stop the search for economy and encode each word. For all messages, the encoder could use a mixture of code numbers and code words, especially since proper names were included in the codebook's vocabulary. Thus the message "Pullman has been appointed Consul-General of the United States at London" with "46248, Bedfellow, Deletery, Masquer" or since London was in the vocabulary, "46248, Bedfellow, Deletery, 38053." Once again, Haswell proudly noted the codebook's design for economy: a message of twelve words was encoded in the equivalent of four words.

From twenty to as high as ninety code word and code number spaces were provided in the codebook at the end of each alphabet letter series, apparently for additional plaintext words or phrases to be added by the State Department and the various embassies as required for particular references.

To construct an even higher communications security fence around the diplomatic dispatches, Haswell added a holocryptic code as an appendix to this codebook several months after the codebook was printed. Entitled *Holocryptic Code, An Appendix to the Cypher of the Department of State*, this system offered fifty rules, each designated by the name of a specific animal, a name that was also found in the plaintext column of the codebook. The sender, after using a particular rule for encoding the message, told the receiver what rule was used by prefixing to the message the code word or code number of the animal named by the particular rule. This prefix was termed the indicator. Indicators could be interpolated into any part of the encoded message: all the message after each indicator would be decoded in accord with the rule represented by that indicator.

Haswell emphasized that encoding required precision and accuracy, and to insure these qualities, the sender, before transmitting, should decode the message after it was encoded in order to provide a better guarantee of correctness. Moreover, the encoder was instructed to write the letters and figures very carefully so the telegrapher could readily read them (the State Department apparently did not introduce typewriters until 1880):[19] this was more important for words than numbers since the telegrapher could misread a letter more readily than a digit.

The fifty rules in the Holocryptic Code provided fascinating variations for further concealment of the encoded dispatch. These rules were designated by fifty animal names ranging from "Ape" through "Deer" and "Pony" to "Zebra." The easiest rule, and the one most attractive to impatient or lethargic diplomats and code clerks, was "Zebra": it simply specified that the message should be encoded and then transmitted without making any change.

The clever designs for these fifty regulations were shaped around three different classes: Route, Addition, and Miscellaneous. The term Route referred to a system of changing the sequence of words in the message by arranging them in an order or route different from that in which they naturally would be written. A key feature of the system consisted in arranging the words up and down in columns. The routing process began when the sender, after encoding the message, divided the words or groups of numbers in the encoded message into as many parts or columns as the particular rule indicated. If there were a remainder after dividing the code words, one word was added to the quotient.

The first pattern within the Route Class called for two columns and provided for routing the encoded words in these columns. For example, in Rule 1, named "Ape," the words of the message were divided into two columns, and the sender began writing codewords of the message at the top of the second column, writing down, and then writing down the first column; in Rule 2, named "Ass," two columns were again used, but the sender wrote the words beginning at the bottom of the second, writing up, and then writing up the first column. The message was then transmitted like a plaintext dispatch, i.e., reading from left to right across the columns. According to the instructions, the Route Class should be used only when there was a minimum of three words in a column. Haswell taught that it would be advisable to change the rule with every message.

The second Route Class pattern contained sixteen variations on routing the encoded words by using three different columns. The following example indicates the complexity of the system. Suppose the following message is to be sent: "A Joint Committee from the Senate and House of Representatives of the United Sates called upon the President and informed him of the organization of the Forty-fourth Congress and their readiness to hear from him."

In code words from the 1876 codebook, this message appears as the following:

AARON LIMPETS CRATES GOSLET SAINT AWARE RECHEATERS STRODE TITCOMB
COCKMAN PLANETARY AWARE KEYHOLE IMPERILED STRODE ORTHOGRAPHY
STRODE GIRLING DECIMATE STROPED PROSECUTOR IMBANKED IMPERILING

If Rule 14, named "Dam," were employed, the coded message was to be arranged in three columns. The rule also required that the sender begin copying the message at the bottom of the second column, then write the message up that column, and then write up the third column, and then write down the first column so the message would appear as follows:

STRODE	STRODE	ORTHOGRAPHY
GIRLIN	RECHEATERS	STRODE
DECIMATE	AWARE	MPERILED
STROPED	SAINT	KEYHOLE
PROSECUTOR	GOSLET	AWARE
IMBANKED	CRATES	PLANETARY
IMPERILING	LIMPETS	COCKMAN
Disgust	AARON	TITCOM

The word "disgust" was added in order to fill the vacant space in the column.

Next the sender referred to the codebook wherein "Dam" was represented by the code word "disinclines," which was the indicator to be transmitted at the beginning of the dispatch. And in the last stage, the message was written from left to right and transmitted as follows:

> DISINCLINES STRODE STRODE ORTHOGRAPHY GIRLING RECHEATERS STRODE DECIMATE AWARE IMPERILED STROPED SAINT KEYHOLE PROSECUTOR GOSLET AWARE IMBANKED CRATES PLANETARY IMPERILING LIMPETS COCKMAN Disgust AARON TITCOMB

The receiver of the dispatch knew from the indicator word "disinclines" that three columns were in the design; and thus he wrote the words in three columns from left to right and then arranged the code words in the order specified in Rule 14. The final step involved using the codebook to find the plaintext words andlor phrases. Besides the sixteen variations of the three-column routing, there were sixteen variations for four-column routes similar to the example noted above.

The Addition Class required the sender to add a specific number to each code number. There are twelve variations ranging from adding a number as low as 33, which had the indicator "Moie," to a high additive of 322, named "Stag." The directions page advised encoding the entire message in code numbers, adding the number specified by the rule, and then transmitting the resulting number, beginning with the indicator. If instead of code numbers, the sender desired to use code words, then he prepared the message in code numbers, added the number specified by the rule, then referred to the codebook and wrote down the code words opposite to this new number, and transmitted the code words.

The Miscellaneous Class included three rules in addition to the "Zebra" rule noted earlier. These variations called for rearranging the code words. The "Tapir" Rule called for substituting the first code word or code number in the message for the second, the second for the first, the third for the fourth, the fourth for the third, etc. The "Tiger" Rule specified the message was transmitted by beginning the message with the last codeword or codenumber and ending it with the first. The third rule in the Miscellaneous Class, named "Wolf," called for letting the first two figures represent the line on the page and the last three numbers represented the page number.

A spelling code completed the last section of the book. Keyed to the first code number in the dispatch, and matching those individual numerals to a special chart. the spelling chart provided a mixed alphabet substitution in a workable though complex manner.

Hamilton Fish, secretary of state, added a new department regulation for this innovative codebook when, in a preface, he ordered that every person authorized to have the book, had to, at the expiration of his term in office or employment, deliver the codebook to the department or a person duly authorized to receive the book.[20] Each person entrusted with the volume was to be be held responsible, and each copy of the book was distinctly numbered. And lastly, each recipient of a copy was required to furnish a receipt for it, and when surrendering the book, obtain a duplicate receipt, one of which was to be forwarded to the department. Never before had State Department codesheets or cipher sheets been numbered; now, for the first time, the complete inventory of department codebooks could be audited and greater security achieved.

It is this codebook, its successor, and the State Department that James Thurber spoofed in a delightful *New Yorker* essay in

1948. Describing his first months as a code clerk in the State Department in 1918, Thurber wrote with much exaggeration that the codebook "had been put together so hastily that the word 'America' was left out and code groups so closely paralleled true readings that 'Lovve' for example, was the symbol for 'love'."[21] In fact, however, the 1876 codebook included a codeword, "Auric" and codenumber, "12641," for "America": the 1899 codebook listed "Beaker" and "14566." The 1876 book did not include code symbols for "love"; its successor in 1899 included it, masked as "leek" and "47768."

Thurber continued his charming account with another code story about his assignment to Europe: "I had been instructed to report to Colonel House at the Hotel Crillon when I got to Paris, but I never saw him. I saw instead an outraged gentleman named Auchincloss, who plainly regarded me as an unsuccessful comic puppet in a crude and inexcusable practical joke. He said bitterly that code clerks had been showing up for days, that Colonel House

Plain message	Number in Vocabulary	Plus 122	Words to be transmitted
A	10000	10422	ABOMINABLE
Joint	36823	36945	LITERALISMS
Committee	20960	21082	CRIMPER
from the	31049	31171	GRANITE
Senate	49221	49343	SAPONIFIES
and	12792	12914	BAILS
House of Reps.	47260	47382	RECOUNTANT
of the	62622	52744	STYLISHLY
United States	54426	54548	TOPIARY
called upon	18992	19114	COIFFURE
the President	44369	44491	PLEASED
and	12792	12914	BAILS
informed	35328	35450	KIRBY
him	33400	33522	IMPROBABLE
of the	52622	52744	STYLISHLY
organization	42039	42161	OUTLET
of the	52622	52744	STYLISHLY
Forty-fourth	30723	30845	GLORIFIES
Congress	21895	22017	DEDITION
and their	52638	52760	SUBDIVIDED
readiness	45995	46117	PROVOKERS
to hear	33217	33339	IMPANELING
from him.	33401	33523	IMPROPED.

As Otter indicates the rule in this example, reference should now be had to the vocabulary, where Otter is represented by the Code Word OTTON, which is the Indicator, and should be prefixed to the message when transmitted. The message will then read as follows:

OTTON ABOMINABLE LITERALISMS CRIMPER GRANITE SAPONIFIES BAILS RECOUNTANT STYLISHLY TOPIARY COIFFURE PLEASED BAILS KIRBY IMPROBABLE STYLISHLY OUTLET STYLISHLY GLORIFIES DEDITION SUBDIVIDED PROVOKERS IMPANELING IMPROPED.

To decipher the above message, reverse the operation.

1876 codebook additive

did not want even one code clerk, let alone twelve or fifteen, and that I was to go on over to the Embassy, where I belonged." And then Thurber explained, "The explanation was, I think, as simple as it was monumental. Several weeks before, the State Department in Washington had received a cablegram from Colonel House in Paris urgently requesting the immediate shipment of twelve or fifteen code clerks to the Crillon, where headquarters for the American Peace Delegation had been set up. It is plain to me now what must have happened. Colonel House's cablegram must have urgently requested the immediate shipment of twelve or fifteen code books, not code clerks. The cipher groups for 'books' and 'clerks' must have been nearly identical, say 'DOGEC' and 'DOGED,' and hence a setup for the telegraphic garble. Thus, if my theory is right, the single letter 'D' sent me to Paris, when I had originally been slated for Berne. Even after thirty years, the power of the minuscule slip of the alphabet gives me a high sense of insecurity. A 'D' for a 'C' sent Colonel House clerks instead of books, and sent me to France instead of Switzerland."[22] Unfortunately, Thurber's engaging and delightful theory did not fit the fact, as seen in the chart below:

Plaintext	1876 Codebook		1899 Codebook	
books	chieftain	18178	columbo	21937
clerks	convexed	20375	cureless	25170

But Thurber appraised an important aspect of the codebooks accurately when he wrote they "were intended to save words and cut telegraph costs."[23]

****

An alarming dispatch from General Horace Porter, the American minister in France, to Secretary John Sherman on the eve of the Spanish-American War told of a new communications threat. Porter enclosed a note from H. R. Newberry, formerly secretary of the American legation at Madrid, and then living in Paris, which stated: "Am sure Madrid Knows State Department cipher."[24] Newberry, whose father was a business associate of General Russell A. Alger, secretary of war, asked Porter to send this information to Alger, and Alger volunteered to be of service at the wish of President McKinley.

A cautious Porter sent a note to Newberry and asked for the source for the cipher information. In a personal message, Newberry replied that upon receiving Porter's request, he sought permission to reveal the name of the person who informed him of the fact that the cipher was known to Madrid officials. Though he could not reveal the name, Newberry added that the informer had not only seen but looked over the State Department book in a Spanish government official's office in Madrid. In addition, he added that it was known that the series of five figures had their order often changed.[25]

Promptly, Porter also warned General Stewart Woodford, the American minister in Madrid in cipher: "Newberry, formerly Secretary of Legation, Madrid, believes Spanish Government has our cipher book. A gentleman he cannot name, says he saw and looked over it, in the hands of a government official."[26] Nevertheless, Woodford continued to employ the 1876 code for his messages to the State Department.

Several weeks after Porter sent his confidential dispatch regarding the apparently compromised cipher, Secretary John Sherman simply replied that he had received the Porter dispatch relative to Spanish-American affairs and offered no comment on the codebook.[27] Woodford became even more troubled by the possibility the codebook had been compromised, for he was very aware of Spanish censorship and activities in the telegraph office. He was also sensitive to the hostile criticism of his accomplishments in Spain as published in *The New York Tribune*. Indeed, only three weeks earlier he had written President McKinley and thanked him for sending the generous telegram of praise: "Coming open in English through the telegraph office where everything is read by the Government censor and communicated to the Ministry, you have also given me the great help of reassuring the Ministry of your personal friendship and confidence. Someone, either at Washington or New York, keeps the Government here posted as to all criticism in the United States of my actions and the *Tribune* dispatches from Washington were embarrassing me."[28]

Troubled Alvey Adee, second assistant secretary, uncertain whether Spanish authorities had the codebook, and reluctant to arouse Spanish suspicions by telegraphing an inquiry to Woodford, since it might be intercepted and read, nevertheless telegraphed Woodford in cipher on April 17: "Have you the printed holocryptic Appendix to the cipher code."[29] And Woodford replied by telegram the next day, "Yes."[30] But by this time, relations between Spain and the United States had deteriorated so badly that substantive cables between Washington and Madrid, in plain text or cipher, even using the holocryptic appendix, were useless. On 21 April, Woodford learned from the Spanish minister that diplomatic relations had been broken, and he left for Paris that afternoon.

Despite the Newberry information, the State Department continued during the Spanish-American War and months afterward to employ the 1876 Red Code for encrypting messages to General Horace Porter in France, Spain's friendly neighbor. Moreover, the department continued to use the codebook as though Porter had never sent the warning dispatch.[31] In May 1899, the department finally switched to the newly published Haswell-designed 1899 Blue Code.

In 1876, America's centennial year, the State Department finally possessed a modern codebook for secret foreign correspondence. The United States, tempted by further overseas expansion and empire, and increasingly cognizant of the dire necessity for secure confidential communications, adopted a better and more practical instrument for masking telegraph and cable dispatches. This new codebook also reflected the flourishing commercial use of codes and ciphers, a practice that flourished in the decades after the invention of the telegraph and transatlantic cable. Public instruments for wire communication required special systems for negotiations and secrecy. Prompt, safe, and accurate official communications could strengthen American economic and political transactions at home and abroad. Technology made code and ciphers essential tools, especially for those engaged in foreign commerce and diplomacy.

Notes

1. Haswell to Fish, Washington, D.C., 8 July 1873, Container 95, Hamilton Fish Papers, Library of Congress.

2. In May 1876, George F. Seward, the American chargé ad interim in Peking, finding no cipher in any of the legation papers, requested one and in late July, Hamilton Fish sent a cipher for temporary use in transmitting confidential telegrams to the State Department. Concerned about economy, he told Seward that another cipher was being prepared and would be sent: "The Cypher should be kept under lock and key, and in consequence of the remoteness of your post, should be used only on occasions when absolutely necessary." Fish to Seward, Washington, D.C., 25 July 1876, *Diplomatic Instructions of the Department of State, 1801-*

1906, China, Microcopy 77, Roll 39, National Archives. The Seward request to Fish, Peking, 3 May 1876, Despatches from U. S. Ministers to China, 1843-1906, Microcopy 92, Roll 41, National Archives.

3. Preface, *The Cipher of the Department of State* (Washington, D.C.: Government Printing Office, 1876).

4. Charles Bright, *Submarine Telegraphs: Their History, Construction, and Working* (London: Crosby Lockwood & Son, 1898), 175.

5. Petition of the New York, Newfoundland, and London Telegraph Co. vs The United States, Filed 25 February

1870, Claim No. 6151, Record Group 59, Microcopy 179, Roll 319, 7, National Archives. When telegraphing figures, telegraphers in each repeating station through which the message was sent were urged to repeat back in order to achieve greater accuracy. This meant double service for the telegraph company, and double charges.

6. Albert B. Chandler, *A New Code or Cipher Specially Designed for Important Private Correspondence by Telegraph and Mail; Applicable As Well to Correspondence by Mail* (Washington, D.C.: Philip Solomons, 1869). Chandler had served with D. Homer Bates, and Charles Tinker in the U.S. War Department Telegraph Office during the Civil War.

7. Bright, *Submarine Telegraphs*, 176.

8. Robert Slater, *Telegraphic Code, to Ensure Secresy [sic] in the Transmission of Telegrams* (London: W. R. Gray, 1870), iii.

9. Ibid., iv.

10. Haswell to Fish, Washington, D.C., 8 July 1873, Container 95, Hamilton Fish Papers, Library of Congress.

Cf. Dumas Malone, ed., "Albert Myer," *Dictionary of American Biography* (New York: Charles Scribner's Sons, 1934), 7:374-375. Also, William H. Plum, *The Military Telegraph During the Civil War in the United States* (New York: Arno Press, 1974), 1:40.

11. "Anson Stager," *Cleveland, Past and Present; Its Representative Men: Comprising Biographical Sketches of Pioneer Settlers and Prominent Citizens with a History of the City* (Cleveland: Maurice Joblin, 1869), 449.

12. Ibid., 449.

13. Plum, *The Military Telegraph During the Civil War in the United States*, 1:44-45.

14. Fish to Porter, Washington, D.C., 20 June 1873, Container 206, Fish Papers, Library of Congress.

15. Ibid.

16. Fish to Chew, Garrison, New York, 10 September 1871, Container 205, Fish Papers, Library of Congress.

17. Chew to Fish, Washington, D.C., 18 September 1871, Container 812, Fish Papers, Library of Congress.

18. Fish to Davis, Garrison, New York, 28 September 1871, Container 106, Fish Papers, Library of Congress: also, Davis to Fish, Washington, D.C., 23 September 1871, Container 82, ibid.: Fish to Davis, Garrison, New York, 29 August 1873, Container 349, ibid.: Davis to Fish, Washington, D.C., 30 September 1871, Container 83, ibid.: Davis to Fish, Washington, D.C., 29 September 1871, ibid.: Fish to Davis, Garrison, New York, 14 June 1872, Container 206, ibid.

19. Typewritten dispatches to France begin 1 July 1898: cf. the correspondence in *Diplomatic Instructions of the Department of State, 1801-1906*, France, Microcopy 77, Roll 64, National Archives.

20. Fish and his staff carefully monitored the cipher security: in a dispatch to Ayres Merrill, the American diplomat in Belgium, Fish observed that a dispatch box containing a cipher had been noted in his predecessor's inventory but not in Merrill's listing, and he requested an explanation. Merrill promptly replied from Brussels that when he took the inventory, there was an iron safe with a combination lock, and not having the combination nor a record of one, he could not open it. However, he proudly reported, he "discovered by chance the 'Open-Sesame' to its contents, and there found the Despatch Box." Cf. Merrill to Fish, Brussels, 2 August 1876, Despatches from U.S. Ministers to Belgium, 1832-1906, Microcopy 193, Roll 15, National Archives: also Fish to Merrill, Washington, D.C., 12 July 1876, *Diplomatic Instructions of the Department of State, 1801-1906*, Microcopy 77, RoB 20, National Archives.

21. James Thurber, "Exhibit X," *The New*

Yorker, 24 (6 March 1948), 26.

22. Ibid., 27.

23. Ibid.,26.

24. Porter to Sherman, Paris, 31 March 1898, Despatches from U.S. Ministers to France, 1789-1906, Microcopy 34, Roll 118, National Archives.

25. Newberry to Porter, Paris, 29 March 1898, in ibid.

26. Porter to Woodford, Paris, 30 March 1898, in ibid.

27. Sherman to Porter, Washington, D.C., April 1898, *Diplomatic Instructions of the Department of State, 1801-1906*, Microcopy 77, Roll 63, National Archives.

28. Woodford to McKinley, Madrid, 6 March 1898, Despatches from U.S. Ministers to Spain, 1792-1906, Microcopy 31, Roll 123, National Archives.

29. Adee to Woodford, Washington, D.C., 17 April 1898, Despatches from U.S. Ministers to Spain, Microcopy 31, Roll 124, National Archives.

30. Woodford to Adee, Madrid, 18 April 1898, ibid.

31. Even the astute Secretary John Hay used the probably compromised code for a telegram to John R. MacArthur at the American embassy in Paris in January 1899. MacArthur, who had been General Woodford's private secretary in Madrid, was informed by Hay that he would be named secretary to the Commission to Investigate Affairs in the Philippine Islands and that the commissioners would have the department cipher.

Chapter 23

U.S. Military Cryptography in the Late Nineteenth Century

David W. Gaddy

The restoration of peace removed whatever impetus to further advancement in cryptology that might have existed in 1865. The vast military machine of the Union contracted swiftly. Other than residual concern about French intentions in Mexico, there was little or no external threat to stimulate thought of such arcane technical matters: The greatly reduced army was stationed in the South as occupation forces, or on remote western posts to cope with the Indians, The wartime U.S. Military Telegraph (USMT) was dissolved soon after Appomattox, and shortly thereafter its wartime rival, the Signal Corps (reduced to a shadow of its wartime strength), finally gained control over electromagnetic telegraphy in the army. Restored to his position as chief signal officer (after a falling out with the secretary of war in 1863), A. J. Myer justified his organization and exercised telegraphy to systematize weather reporting to Washington, and thereby laid the foundation for the Weather Bureau.[1] His few men retained a scientific and practical interest in electricity and were quick to seize upon Alexander Graham Bell's telephone in the mid-1870s.

For military cryptography, Stager's Civil War route transposition system continued to be the War Department's telegraphic cryptosystem in the 1870s, until the promulgation of a new code (dated 1885) in 1886. Army signal chief A.J. Myer, who had clashed with the strong-minded wartime secretary, Edwin M. Stanton, over control of wire telegraphy, was reinstated in 1866 as chief signal officer.[2] With the disbanding of the rival wartime U.S. Military Telegraph (and a new secretary), electromagnetic telegraphy came within the purview of the Signal Corps.[3] Stager's wartime systems were replaced with a new family of route transposition systems, little changed from those they replaced. Indicators ("commencement words") showed the dimensions of the matrix (rows, columns); code words ("arbitraries"), usually with one or more alternates, concealed names, numbers, places, words, and phrases; fillers or nulls were prescribed; and a substitution cipher was provided to encipher names or words not contained in the vocabulary of the system.

The Army's role in the '70s was that of occupying power in the former Confederate States and protector-pacifier in the West. What concern there was for confidentiality aimed mainly at former rebel telegraph operators and zealous newsmen who had learned to copy "by ear." It was generally adequate to simply distort the plain text, and the Stager system did quite well.

In the 1870s and '80s, both the Army and Navy (dwarfed now by the navies of England and France) sought to build their professional ranks through advanced education, the study of their own history and the experience of other nations, and through the cultivation of professional literature. Aging veterans of the Civil War were recording their service and refighting their battles in public, aided by fraternal activities of veterans' organizations, and, in this manner, some record of the USMT, the Signal Corps, and their Confederate counterparts began to emerge. Reflecting a growing interest in foreign counterparts, the Navy established an Office of Naval Intelligence in 1882, followed shortly thereafter by the Army's Military Information Division. To improve the quality of military education, the Navy established a post-academy War College in 1884, followed by the Army a few years later.

The Stager route transposition finally passed out of official use in 1886 with the publication of the Army's first "new model" telegraphic code. Dated 1885, but distributed the following year, the *Telegraphic Code to Insure Secrecy in the Transmission of Telegrams*, compiled under the direction of Lieutenant Colonel J.F. Gregory, an aide to Lieutenant General W.T. Sherman, the commanding general, was a one-part code of 25,000 entries, little changed from a commercial publication of the same name by Robert Slater. Carried in five-digit code groups, it depended upon superencipherment (separately prescribed) for its security. (It was indicative of the state to which cryptology had fallen in the Signal Corps that they were not involved in this issuance.)

America's first venture into foreign wars since the War with Mexico (the Spanish-American War) saw forces committed to Cuba and the Philippines and exposed the poor state of readiness of the American military. In *The American Black Chamber* (1931), Herbert O. Yardley (who was to create America's first military cryptologic organization, The Cipher Bureau of Military Intelligence (MI-8) in 1917) recounts the story of an Army veteran of that struggle saying that a single, constant additive of "1898," the year of the war, was used as the "superencipherment" of the old 1885-86 telegraph code, still in use during the Spanish-American War.[4]

The Navy's entry into "modern" cryptography was its "Secret Code" of 1885, a massive tome, with two supplemental volumes, that also involved super-encipherment of code. Friedman, in his "Lectures," uses the historic Theodore Roosevelt-to-Admiral Dewey message of 25 February 1898 to illustrate the working of this cumbersome system, which continued to serve the Navy until 1915.[5]

Although the telephone and wire telegraph were employed, the Signal Corps still used Myer's Civil War era visual flag system, along with the heliograph (introduced out west in the decade after the Civil War ended) and carrier pigeons. With the exception of cable interception of Spanish government communications and some simple "tactical cryptanalysis efforts" in the field during the subsequent Filipino Insurrection, there seems to be little else to say about cryptology or signals intelligence during this period.

Under General Orders 9 of 16 January 1898, the War Department directed that a new War Department telegraph code be prepared by the Chief Signal Officer (scientist-soldier A.W. Greely) to replace the 1885 code. Greely's labors were overtaken by the Spanish-American War, and in 1899-1900 a "Preliminary" code (one-part, unenciphered) was issued under the authority of the reformist secretary of war, Elihu Root, with the 1885 Gregory code to be kept in reserve for "special cases." A "Cipher" (actually another code, one-part, developed by W.H. Ainsworth and W.G. Spottswood – the clerk who compiled the 1885 code) was introduced in 1902, and finally Greely's enciphered code was issued on 1 February 1906.

Nevertheless, by the turn of the century military and scientific inquiry was pressing ahead. International Morse had displaced American Morse (except on wire lines), and the Myer "General Service" (army-navy) code had adopted International Morse equivalents. Interest in wireless telegraphy in the 1890s, followed shortly thereafter by wireless telephony, ushered in the era of radio. At the army service schools in Kansas, technical conferences explored the implications of modern warfare: among the participants were a future chief signal officer of the Army, Joseph O. Mauborgne, and the author of what historian David Kahn considered the first significant publication on cryptography in English since Dr. William Blair wrote in the early 1800s, Parker Hitt. Hitt's 1916 *Manual of Military Cryptography* coincided with preparation for American entry into the European war. That same year the Signal Corps deployed radio interception and high frequency radio direction-finding units along the border with Mexico. By this time, "communication" and "message" had come to replace "signal" in U.S. Army usage, and "wireless" had given way to the French-derived term, "radio." A new era had dawned, both for communication and for cryptology, the era of radio: new challenges for cryptography and communication security; new opportunities for "radio intelligence." No more the intercepted courier, the observed visual signals, the occasional wiretap. With electronics came modern cryptology.

Notes

1. Myer's diversion to meteorology prompted General Sherman, in an 1874 budget defense, to term the "Signal Service" more scientists than soldiers and to suggest that their interests in the weather be turned over to the Smithsonian. He growled that the soldier had no interest in the weather – he had to endure it and fight in it, come what may.

2. Myer may have adopted the Blair essay on "Cipher" and had it reprinted as a text and reference. See chapter 14.

3. *A Manual of Military Telegraphy for the Signal Service, United States Army* – prepared under the direction of the chief signal officer of the army – was published by the Government Printing Office in 1872.

4. *Friedman Legacy,* 131. In his "Lectures," Friedman abandoned his customary caution to accept the Yardley story and speculate that the numbers "777" appearing in ink inside the cover of the copy of the 1885-86 army codebook he owned was a constant additive. In all probability it was the owner's copy number, assigned for security accounting purposes.

5. See *Friedman Legacy,* 128 and 135-138.

Chapter 24

The Blue Code of the Department of State, 1899

In January 1898, four years after his retirement from the State Department, John Haswell wrote a very realistic and discerning letter on communications security to the elderly secretary of state, John Sherman. Explaining that the present code of the State Department had been in use for almost twenty-five years, he warned the former Ohio senator, "It is well known that the European telegraphic system is an adjunct to the Postal arrangement of each country." Sensitive to European espionage practices, Haswell continued, "Every telegraphic cipher message entering into or passing through a country is sent to the Foreign Office and referred to the Bureau of Ciphers, where it is recorded and efforts are made to translate it. This is the rule adopted by the powers for all government communications."

Alert to the codebreaking activities of foreign governments, Haswell continued in his letter to caution the secretary, as John Bigelow had sought to convince William Seward earlier, that the interception of foreign dispatches was commonplace. Haswell explained, "In France the Cipher Bureau, attached to the Ministry for Foreign Affairs, employs from fifteen to twenty clerks whose sole duty is to classify all such messages with a view to obtain as complete a knowledge of the codes of the various countries as possible. It cannot be questioned that however complete a code may be, in the course of twenty-five years the trained efforts of these Bureaus would have given them at least a very good general knowledge of the cipher of the Department."[1]

In late 1897 and early 1898, threats and rumors of war with Spain over the issue of Cuban independence continued to shape the headlines and crowd the front pages of American daily newspapers, especially the "Yellow Press" along the eastern seaboard. Undoubtedly, war fever and the danger of foreign intervention influenced Haswell as he sought greater security for American dispatches. He wrote that since European Black Chambers operated freely, and probably possessed copies of the State Department's 1876 Cipher, "it would seem good policy, after a service of twenty-five years to establish a new and improved system of telegraphic communication between the Department and its Diplomatic Agents."[2]

Eager to prepare a new code, Haswell also argued that all cablegrams should be sent in code, no matter how unimportant: it is crucial that secrecy be maintained. Furthermore, the State Department code should be completely changed so that cryptographic information in the hands of foreign governments would not apply to the new code. For economy, the sentences in the old code would be changed and many new ones introduced, thus making it different from the old one, and also reducing the expenses of cable transmission. And, finally, he suggested no specific stipend for himself for the preparing the codebook, leaving that decision to the department.

On 24 March 1898, during the same hectic week that the war with Spain was declared, Haswell finally received an answer to his proposal. Because of his experience, the needs of the department (weeks earlier, General Horace Porter, the American minister to France, had alerted the State Department that the Spanish government probably had the department's 1876 code) and also his earlier codebook, Haswell received the desired assignment from Sherman. The secretary had earlier endorsed the idea for better communications security in a January letter. And Sherman needlessly warned the cautious Haswell: "Nor will it be necessary to call your attention to the confidential character of the undertaking. The Department therefore confidently believes that you will exercise the greatest care in order to prevent a knowledge of its preparation, or any of the material from getting into the possession of unauthorized parties, whereby its usefulness might be impaired. Your compensation will be $3,000."[3]

Within a year, Haswell's codebook was submitted to the department and accepted by Sherman's successor, the experienced secretary of state, John Hay. Instructions for its use emphasized that the volume, each one numbered, was to be considered as special property entrusted to heads of missions and these officers were to deliver copies to their permanent or ad interim successors in office and require a special receipt in duplicate for the book.

Anxious about developing better security measures, John Hay added that if a head of mission left his post on leave or otherwise, he had to return the book under seal by safe and speedy conveyance to the State Department or place the volume in the temporary custody of some diplomatic or consular officer of the United States. Here again, duplicate receipts must be filed. If no officer was available, then the volume in a sealed envelope was to be delivered to a United States naval officer with the request that he deliver it safely and promptly to the department. The codebook must never be placed in the possession of the archives custodian. And, finally, only the head of mission or the first secretary of the embassy or legation could have access to *The Cipher of the Department of State*.[4]

The extensive codebook, which became known as the *Blue Code*, remained the State Department's primary codebook for the next eleven years. Its title page is identical to the 1876 codebook except that John Haswell is listed as the author. This similarity between the two codebooks probably resulted in the former book coming to be identified by the color of its cover, red, and the latter, blue. Less than ten months after completing this second generation State Department codebook, John Haswell died on 14 November 1899.[5]

No. 26 DEPARTMENT OF STATE.

WASHINGTON, Oct 8, 1900

RECEIVED OF THE CHIEF CLERK OF THE DEPARTMENT OF STATE, CIPHER NO. 26 WITH HOLOCRYPTIC CODE WHICH I WILL CARE FOR AND USE IN ACCORDANCE WITH THE INSTRUCTIONS OF THE DEPARTMENT OF STATE CONTAINED THEREIN.

DUPLICATE

Wilbur J. Carr.

Receipt for 1899 code

The 1899 edition resembles the earlier codebook in design and format. Whereas the first book left blank spaces for adding additional plaintext words and/or phrases at the end of each alphabetical code group, the later one left the spaces at the beginning of each code group. Thus, the word A was masked by the code number "10000" and the code word "Aaron" in the first tome; in the second book, the respective codes were "10425" and "Accessibly." Also the second book expanded the plaintext words and phrases beginning with A to 8,650 items, thus 2,582 more than the 1876 book. The 1899 book numbered almost 1,500 pages compared to 1,200 in Haswell's first volume. For both books, and especially the latter, Haswell selected plaintext words and phrases that were common in American diplomatic correspondence.

Haswell explains in the instructions section that the International Telegraphic Conference held in Paris determined that ten letters or four figures constituted one word. Each codeword in this book consisted of ten or fewer letters. However, code numbers contained five figures, and therefore each group of figures constituted two words and would be assessed double charges. Though lacking in economy, the code numbers had the advantage over the word system because of its applicability to all languages that contain numerals.

Therefore, Chinese, French, German, Italian, Japanese, Russian, and Turkish operators who often committed critical errors in transmitting word messages were familiar with figures, and thus transmitted them with accuracy. However, more recently, these telegraphers became more skilled, and especially those personnel in charge of international transmissions achieved a much higher level of accuracy in transmitting and receiving foreign language messages. Because of this and the additional fact that fewer errors were being made during transmission of the large volume of commercial messages that use code words, Haswell wrote that "it seems advisable that the economical advantages offered by the Code Word system be availed of, and its use in preference to the use of the figure system is therefore recommended."[6] Unreasonable concerns about economy again drove the State Department communications security system.

Code word P	Code No 609	Message or true reading
Promotes	00	Russia
Promoting	01	Agreement between Russia and
Promotion	02	Agreement with Russia
Promotions	03	Ambassador from Russia
Promotive	04	Ambassador of Russia
Prompt	05	Ambassador to Russia
Prompted	06	And Russia
Prompter	07	Army of Russia
Prompters	08	Authorities of Russia
Promptest	09	Authority of Russia
Prompting	10	By Russia
Promptings	11	Cabinet of Russia
Promptly	12	Charge d'affaires of Russia
Promptness	13	Commerce of Russia
Prompts	14	Consul of Russia
Promulgate	15	Consul-general of Russia
Pronate	16	Consuls of Russia
Pronation	17	Convention between Russia and
Pronator	18	Convention with Russia
Prone	19	Czar of Russia
Pronely	20	Embassy of Russia
Proneness	21	Emperor of Russia
Prong	22	Empire of Russia
Pronged	23	Empress of Russia
Pronghorn	24	Flag of Russia
Prongs	25	Forces of Russia
Pronity	26	From Russia
Pronoun	27	From the Government of Russia
Pronounce	28	Government of Russia
Pronounced	29	Head of the Government (by whatever title)
Pronouncer	30	Her Majesty the Empress of Russia
Pronounces	31	His Majesty the Emperor of Russia
Pronouns	32	Imperial Government of Russia
Pronubial	33	In Russia
Pronuncial	34	Legation of Russia
Proof	35	Minister for foreign affairs of Russia
Proofless	36	Minister for foreign affairs of Russia (by name)
Proofs	37	Minister from Russia
Prop	38	Minister of Russia
Propagable	39	Minister of Russia at
Propaganda	40	Minister to Russia
Propagate	41	Naval vessel of Russia
Propagated	42	Naval vessels of Russia
Propagates	43	Navy of Russia
Propagator	44	Of Russia
Propel	45	People of Russia
Propelled	46	Policy of Russia
Propeller	47	Possessions of Russia
Propellers	48	Secretary of embassy of Russia
Propelling	49	Secretary of legation of Russia

1899 Blue Code Sheet

Code word P	Code No 609	Message or true reading
		Russia - Continued
Propels	50	Subject of Russia
Propend	51	Subjects of Russia
Propended	52	The Czar of Russia
Propendent	53	The Emperor of Russia
Propending	54	The Empire of Russia
Propends	55	The Empress of Russia
Propense	56	The Government of Russia
Propensely	57	The trade of Russia
Propension	58	To Russia
Propensity	59	To the Government of Russia
Proper	60	Treaty between Russia and
Properate	61	Treaty with Russia
Properly	62	Vessel of Russia
Properness	63	Vessels of Russia
Properties	64	War vessel of Russia
Property	65	War vessels of Russia
Prophesied	66	With Russia
Prophesies	67	With the Government of Russia
Prophesy	68	Russian
Prophet	69	And the Russian
Prophetess	70	And the Russian Government
Prophetic	71	From the Russian Government
Prophetize	72	Russian agreement
Prophets	73	Russian Ambassador
Propine	74	Russian Ambassador (by name)
Propined	75	Russian army
Propines	76	Russian authorities
Propining	77	Russian authority
Propitious	78	Russian cabinet
Proplastic	79	Russian capital
Propodial	80	Russian charge d'affaires
Propone	81	Russian commerce
Proponed	82	Russian company
Proponent	83	Russian consul
Propones	84	Russian consul-genera,l
Proponing	85	Russian consuls
Proportion	86	Russian convention
Proposal	87	Russian Czar
Proposals	88	Russian embassy
Propose	89	Russian Emperor
Proposed	90	Russian Empire
Proposer	91	Russian Empress
Proposers	92	Russian flag
Proposes	93	Russian forces
Proposing	94	Russian funds
Propound	95	Russian Government
Propounded	96	Russian lagation
Propounder	97	Russian minister
Propounds	98	Russian minister (by name)
Propped	99	Russian minister at

1899 Blue Code Sheet

Sometime later, probably several years, a revised instruction was added to the codebook. Because of telegraph tariff revisions, the new guideline specified there was now no added expense in transmitting five-digit code numbers. Thus, the department recommended that code numbers be used because figures are similar in almost all languages, and fewer mistakes result when code number groups are transmitted.

A Spelling Code, to be used for proper names and certain places, in the 1899 codebook also resembled the first edition. The following example indicates how the fairly complicated procedure had to be followed. If the code clerk wished to encode the phrase "John Hay, our former Ambassador to Great Britain, is now Secretary of State," the plan required the clerk first encipher the name, John Hay. He had to find the number corresponding to the first code word in the dispatch. Thus, the first plaintext word after the name was our, and the codeword for it was "MUSLINET" and code number, "52940." This codenumber had to be written and partially repeated over the name John Hay.

5	2	9	4	0	5	2
J	O	H	N	H	A	Y

Then the clerk had to substitute for the plaintext letters with the distorted alphabets specified by the numbers. Table No. 1 at the back of the book was used for the first five letters. Thus, for plaintext J, the fifth distorted alphabet provided the letter T. In distorted alphabet No.2, S was found for plaintext O. Alphabet No. 9 gave an A for an H: alphabet No. 4 showed "I" for N: alphabet No. 0 listed "N" for H. The clerk then was required to turn to Table No. 2 to complete the masking of the last two plaintext letters that followed the same design as Table 1: distorted alphabet No. 5 showed a "B" for plaintext A, and alphabet No. 2 listed "F" for Y. (See chart below.)

For decoding the spelling code letters, the process was reversed: the numbers of the Key Group should be placed under the spelling cipher letters, and the numbers indicated the particular distorted alphabet from which these letters were selected. The letters in the regularly arranged alphabet over the cipher letters provided the plaintext letters.

Also prepared by Haswell to accompany the 1899 codebook was the sixteen-page *Holocryptic Code, An Appendix to the Cipher of the Department of State,* also published by the Government Printing Office in 1899. The thin hardcover book resembled the holocryptic codebook devised by him a quarter of century earlier. The new edition, devised for "greater security," had seventy-five rules, twenty-five more than the earlier one. These rules described additional secret systems for veiling the messages: 1) thirty-four rules for changing the message route

KeyGroup	5	2	9	4	0	5	2
Plain text	J	O	H	N	H	A	Y
Cipher	T	S	A	I	N	B	F

Using code words, the message would read as follows:

TSAINBF	MUSLINET	GATERING	GRINDLE
JohnHay	our	former	Ambassador to Great Britain
MISTRUST	RAMAGE		
is now	Secretary of State		

If code numbers were used for the message, it would read as follows:

TSAINBF 52940 38086 39733 51696 62018

into two, three, and four columns; 2) thirteen rules for adding specific numbers to the codenumbers; 3) twelve rules for subtracting numbers from the code numbers; and 4) a series of thirteen miscellaneous rules such as substituting the first code word or code number for the second, the second for the first, the third for the fourth, and fourth for the third, etc. Haswell's 1876 edition did not include subtraction rules.

All the rules were designated by short familiar words, primarily the names of animals such as "cow" and "bison" and "dog." All these names could be found in the codebook's plain text, and thus the sender, when employing a particular rule for encrypting the message, revealed to the recipient that rule had been used by prefixing to the message the code number or code word of the plaintext rule. This code number or code word was termed the "indicator."

The instructions also specified that indicators could be interpolated into any part of the encrypted message: all the encrypted message after the indicator had to be translated by the recipient in accord with the rule specified by the indicator. Under the general remarks, Haswell again emphasized that the location of the indicator code word or code number should never change; it invariably had to be the first item. With genuine enthusiasm, Haswell concluded his instructions in the *Holocryptic Code* book with the following statement: "It is believed by the compiler of these Rules that their application to the Code Vocabulary will secure the utmost secrecy and render cipher messages wholly inscrutable."[7] And for the harried code clerks, he added, "Facility in the use of these Rules will come with practice and will render the extra time needed for their application quite inappreciable in comparison with the object accomplished."

How did John Haswell evaluate his new codebook in describing it to the American public; what were its principal characteristics and merits? Here is his unique appraisal, published in a popular monthly magazine twelve years after his death (he did not note in the article that he designed the book): "The cipher of the Department of State is the most modern of all in the service of the Government. It embraces the valuable features of its predecessors and the merits of the latest inventions."[8] Sensitive to the cipher's considerable size, Haswell explained, "Being used for every species of diplomatic correspondence, it is necessarily copious and unrestricted in its capabilities, but at the same time it is economic in its terms of expression."

Gratified that the design offered State Department code clerks a readily convenient mask, Haswell wrote proudly, "It is simple and speedy in its operation, but so ingenious as to secure absolute secrecy." He concluded his appraisal with a somewhat mysterious sentence that must have been calculated, for security reasons, to veil the nature of the secret system. He wrote of the "key" to the cipher and did not explain that one could simply unlock the State Department cipher by obtaining the codebook. Instead he wrote, "The construction of this cipher, like many ingenious devices whose operations appear simple to the eye but are difficult to explain in writing, would actually require the key to be furnished for the purpose of an intelligible description of it."

A significant addendum to the Blue Codebook was made in September 1900 by Alvey A. Adee, acting secretary of state, who was then in his thirty-first year of service to the State Department (this Shakespearean scholar and perceptive administrator served almost fifty-five years, leaving the department only a week before his death in 1924 at the age of eighty-one).[9] This annex described the precise method for encoding the time and date that the telegram was sent: its purpose - economy: "giving the information at the smallest possible expense."[10] Numbers and letters shaped the code:

0	1	2	3	4	5	6	7	8	9	10	11	12
O	A	B	C	D	E	F	G	H	J	K	L	M

The letters were used for hiding the month and the hour of the day: the twelve numbers signified the month and hour. The month was to be named first, and was represented by a letter in alphabetical order: *January*, being the first month, was represented by "A," *February*, "B"; the code continued on to *December*, which was "M." The letter "I" was not part of the code since it often was confused with "J."

Each day in the month was represented with two letters: "OA" represented the first day of the month; "OK" for the tenth day; and "AC" for the thirteenth day. An important rule stated that the letter "O" was placed in front of any letter representing any one of the first twelve numbers so that the date was always represented by two letters, preventing a possible mistake if one of the letters was dropped during transmission. This mistake would change the date from the end of the month to the beginning. Finally, the month-day-hour group, which was always five letters in length, gave the hour of the day or night represented by a letter that was followed by the letter "A" for A. M., or "F" for P. M,: "N" for midnight; "M" for noon.

A dispatch filed at 7 P.M. on 3 September would begin with "JOCGP," since September is the ninth month; "OC" for the third day; and "GP" for the seventh hour of the afternoon. Another example: a telegram sent at noon on 28 February would begin with "BBHMM" for the second month, 28th day, and twelve noon.

Again, these instructions emphasized economy by stating that cable and telegraph companies agreed that a five-letter group was one word.

Secretary of State John Hay sent the very first copy of the codebook and holocryptic code by way of Paris to Bellamy Storer, the new envoy extraordinary and minister plenipotentiary to Spain in May 1899. "In the diplomatic pouch, securely sealed in a wooden box" and addressed to General Horace Porter, the American minister in Paris, the secret book and appendix were to be given to Storer when he called at the embassy on the way to Madrid.[11] Storer, a fifty-one-year-old lawyer and former Ohio Republican congressman, had just completed two quiet years as minister to Belgium, an appointment that came as a reward for his support of William McKinley in the 1896 presidential election.

Storer, faced with negotiations in Spain on such difficult postwar problems as the status of church property in the Philippine Islands, the return of Spanish prisoners of war held in those islands, and also the release of Cuban political leaders in Spain, might well have needed the innovative codebook. Hay told him about the new code and instructed him to deposit it in the archives of the legation in Madrid after signing the original receipt, which had to be returned to the State Department. Aware of European espionage skills because of his prior service as secretary of the American legation in Paris under John Bigelow, Hay added, "The greatest care should be taken by you to insure the safety and preserve the secrecy of the cipher."[12]

Despite all the State Department warnings about exercising special security precautions, the new codebook was stolen in St. Petersburg sometime before mid-1905 and the appointment of the new American minister, George von Lengerke Meyer, to that post.[13] Although President Theodore Roosevelt angrily denounced the lack of proper security procedures that resulted in loss of the codebook, the State Department failed to publish a new codebook. Rather, this 1899 volume and appendix would remain the primary State Department encryption system until 1910 when the *Green Cipher: The Cipher of the Department of State appeared.*[14] And despite the book's loss a decade earlier, President Woodrow Wilson and his trusted advisor, Colonel Edward House, would use the 1899 book for their private correspondence in the early years of World War I.

Notes

1. John Haswell to John Sherman, Washington, D.C., 26 January 1898. Copy in author's possession. In the printed directions for the new codebook, Haswell did not mention intercept practices as a reason for the new book; rather, he simply noted "Changed conditions in general and reasons of a special character have made it desirable for the Department to have a new and enlarged Code": cf. *The Cipher of the Department of State* (Washington, D.C.: Government Printing Office, 1899), 5.

2. Ibid.

3. John Sherman to John Haswell, Washington, D.C., 24 March 1898. Copy in author's possession.

4. John Hay instructions sent with letter of Alvey Adee to John Haswell, Washington, D.C., 20 March 1899. Copy in author's possession.

5. *The Times Union* (Albany, N.Y.), 15 November 1899.

6. *The Cipher of the Department of State* (1899), 7.

7. Haswell, *Holocryptic Code*, 16.

8. John H. Haswell, "Secret Writing: The Ciphers of the Ancients, and Some of Those in Modern Use," *The Century Illustrated Monthly Magazine*, 85 (November 1912), 92.

9. A dedicated worker, Adee slept on a cot in his office so that he could be available to decode the latest messages from Madrid and Havana during the growing crisis with Spain in the spring of 1898: cf. John A. De Novo, "The Enigmatic Alvey A. Adee and American Foreign Relations, 1870-1924," *Prologue* (Summer 1975), 77.

10. This page appearing above Adee's name and dated 14 September 1900, is bound into the Blue Code book just before the title page.

11. Hay to Porter, Washington, D.C., 19 May 1899. *Diplomatic Instructions of the Department of State, 1801-1906*, Microcopy 77, Roll 64, National Archives.

12. Hay to Storer, Washington, D.C., 19 May 1899, Instructions, Microcopy 77, Roll 150. National Archives.

13. George von Lengerke Meyer to Theodore Roosevelt, St. Petersburg, 5 July 1905, in The Presidential Papers of Theodore Roosevelt, Series 1, Roll 56: also Roosevelt to Meyer, Oyster Bay, 18 July 1905, in Elting E. Morison, ed., *The Letters of Theodore Roosevelt* (Cambridge: Harvard University Press, 1951), 4:127.

14. Ralph E. Weber, *United States Diplomatic Codes and Ciphers, 1775-1938*, 246.

Bibliography

Ameringer, Charles D. *U.S. Foreign Intelligence: The Secret Side of American History*. Lexington, Massachusetts: Lexington Books, 1990.

Bakeless, John E. *Spies of the Confederacy*. Philadelphia: J.B. Lippincott Company, 1970.

_______________ *Turncoats, Traitors and Heroes*. Philadelphia: Lippincott, 1959.

Bates, David H. *Lincoln in the Telegraph Office: Recollections of the United States Telegraph Corps During the Civil War*. New York: The Century Co., 1907.

Bemis, Samuel Flagg. *The Diplomacy of the American Revolution*. New York and London: D. Appleton-Century Co., 1935.

Bidwell, Bruce W. *History of the Military Intelligence Division, Department of the Army General Staff*. 8 vols. Washington, D.C.: Department of the Army, 1959-1961.

Boyd, Julian P.; Cullen, Charles T.; Catanzariti, John, eds. *The Papers of Thomas Jefferson*. 24 vols. Princeton: Princeton University Press, 1950-1991.

Bright, Charles. *Submarine Telegraphs: Their History, Construction, and Working*. London: Crosby Lockwood & Son, 1898.

Burnett, Edmund C., ed. *Letters of Members of the Continental Congress*. 7 vols. Washington: The Carnegie Institution, 1921-1936.

Butterfield, Lyman H., and Friedlaender, Marc, eds. *Adams Family Correspondence*. 4 vols. Cambridge: Harvard University Press, 1963-1973.

Creason, Steve. "Cryptology Played Important Role in Our Nation's History." *NSA Newsletter* (reprinted from *The Hallmark*), June 1976, 5.

Currey, Cecil B. Code Number 72/ *Ben Franklin: Patriot or Spy?* Englewood Cliffs, New Jersey: Prentice-Hall, 1972.

Davis, Matthew L., ed. *Memoirs of Aaron Burr*. New York: Charles Scribner's Sons, 1966.

Deacon, Richard. *A History of the British Secret Service*. New York: Taplinger Publishing Co., 1969.

Dorwart, Jeffery M. *The Office of Naval Intelligence: The Birth of America's First Intelligence Agency 1865-1918*. Annapolis, Maryland: Naval Institute Press, 1979.

Ellis, Kenneth. *The Post Office in the Eighteenth Century: A Study in Administrative History*. London: Oxford University Press, 1958.

Fitzpatrick, John C., ed. *Diaries of George Washington from the Original Manuscript Sources, 1745-1799*. 39 vols. Washington: Government Printing Office, 1931-1944.

Flick, Alexander Clarence. *Samuel Jones Tilden: A Study in Political Sagacity*. Port Washington, New York: Kennikat Press, 1963.

Freeman, Douglas Southall. *George Washington: A Biography*. 4 vols. New York: Charles Scribner's Sons, 1951, Vol. 3, 541-55.

The Friedman Legacy. Ft. George G. Meade, Maryland, Center for Cryptologic History, 1992.

Friedman, William F., and Lambros D. Callimahos. *Military Cryptanalytics - Part I*. SRH-273, 1956, 30.

Friedman, William F., *The Friedman Lectures*. SRH-004, 1963, 37.

Fuller, Wayne E. *The American Mail*. Chicago: University of Chicago Press, 1972.

Green, James R. *The First Sixty Years of*

the Office of Naval Intelligence. Washington: American University, 1963.

Hall, Charles Swain. *Benjamin Tallmadge: Revolutionary Soldier and American Businessman*. New York, 1943. Reprint., New York: AMS Press, Inc. 1966.

Haswell, John. *The Cipher of the Department of State*. Washington, D.C.: Government Printing Office, 1876.

_______________ *The Cipher of the Department of State*. Washington, D.C.: Government Printing Office, 1899.

Haworth, Paul Leland. *The Hayes-Tilden Disputed Presidential Election of 1876*. Cleveland: The Burrows Brothers Company, 1906.

Herring, James M., and Gross, Gerald C. *Telecommunications: Economics and Regulation*. New York and London: McGraw Hill, 1936.

Hitt, Parker. *Manual for the Solution of Military Ciphers*. Fort Leavenworth, Kansas: Press of the Army Service Schools, 1916.

Hopkins, James F.; Hargreaves, Mary W.; Seager, Robert II; Hay, Melba, eds. *The Papers of Henry Clay*. 10 vols. Lexington: University of Kentucky Press, 1959-1991.

Howeth, Linwood S. *History of Communications - Electronics in the United States Navy*. Washington, D.C.: G.P.O., 1963.

Hutchinson, William T.; Rachal, William M.; Rutland, Robert A.; Hobson, Charles F.; Mason, Thomas A.; Sisson, Jeanne K.; Brugger, Robert J.; Crout, Robert R.; Dowdy, Dru; Stagg, John C.; Mattern, David B.; Cross, Jeanne K.; Perdue, Susan H., eds. *The Papers of James Madison*. 18 vols. Chicago: University of Chicago Press, 1962-1977 (vols. 1-10). Charlottesville: University of Virginia Press, 1977-1992 (vols. 11-18).

Kahn, David. *The Codebreakers: The Story of Secret Writing*, New York: Macmillan, 1967.

_______________.*Kahn on Codes*. New York: Macmillan, 1983.

Lomask, Milton. *Aaron Burr, The Conspiracy and Years of Exile, 1805-1836*. New York: Farrar, Straus, Giroux, 1982.

Marshall, Max L., ed. *The Story of the U.S. Army Signal Corps*. New York: Franklin Watts, Inc., 1965.

O'Toole, G.J.A. *Encyclopedia of American Intelligence and Espionage*. New York: Facts on File, 1988.

_______________ *Honorable Treachery: A History of U.S. Intelligence, Espionage, and Covert Action from the American Revolution to the CIA*. New York: The Atlantic Monthly Press, 1991.

_______________ *The Spanish War: An American Epic, 1898*. New York: Norton, 1984.

Pennypacker, Morton. *George Washington's Spies on Long Island and in New York*. Brooklyn: Long Island Historical Society, 1939.

Plum, William R. *The Military Telegraph in the Civil War in the United States*. 2 vols. Chicago, 1882. Reprint (2 vols. in 1). New York: Arno Press, 1974.

Porter, Bernard. *Plots and Paranoia: A History of Political Espionage in Britain 1790- 1988*. London: Unwin Hyman, 1989.

Powe, Marc B., and Wilson, Edward E. *The Evolution of American Military Intelligence*. Fort Huachuca, Arizona: The United States Army Intelligence Center, 1973.

Pratt, Fletcher. *Secret and Urgent: The Story of Codes and Ciphers*. Garden City, New York: Blue Ribbon Books, 1942.

Rees's Clocks, *Watches and Chronometers, 1819-1820: A Selection from the Cyclopaedia of Universal Dictionary of Arts, Sciences and Literature*. Rutland, Vermont: C.E. Tuttle Co., 1970.

Rowan, Richard W., and Deindorfer, Robert G. *Secret Service: 33 Centuries of Espionage*. New York: Hawthorn Books, 1967.

Slater, Robert. *Telegraphic Code, to Ensure Secresy [sic] in the Transmission of Telegrams*. London: W. R. Gray, 1870.

Stern, Philip. *Secret Missions of the Civil War*. Chicago: Rand McNally and Co., 1959.

Stevens, Benjamin Franklin. *B.F. Steven's Facsimiles of Manuscripts in European Archives Relating to American 1773-1783*. 25 vols. London, 1898. Reprint., New York: AMS Press, 1970.

Tallmadge, Benjamin. *Memoirs of Colonel Benjamin Tallmadge*. New York, 1858. Reprint., New York: Arno Press, 1968.

Taylor, John M. *William Henry Seward: Lincoln's Right Hand*. New York: Harper Collins, 1991.

Thompson, James W., and Padover, Saul K. *Secret Diplomacy: Espionage and Cryptography, 1500-1815*. New York: Frederick Ungar, 1963.

Tidwell, William A.; Hall, James o.; Gaddy, David W. *Come Retribution: The Confederate Secret Service & the Assassination of Lincoln*. Jackson, Mississippi: University Press of Mississippi, 1988.

Van Doren, Carl C. *Secret History of the American Revolution*. New York: Viking Press, 1968.

Wallace, Willard M. *Appeal to Arms: A Military History of the American Revolution*. New York: Quadrangle/The New York Times Book Co., 1975, 1-52.

Wharton, Francis, ed. *The Revolutionary Diplomatic Correspondence of the United States*. 6 vols. Washington, D.C.: Government Printing Office, 1889.

Weber, Ralph E. *United States Diplomatic Codes and Ciphers, 1775-1938*. Chicago: Precedent, 1979.

Wriston, Henry Merritt. *Executive Agents in American Foreign Relations*. Baltimore: Johns Hopkins Press, 1929.

www.ingramcontent.com/pod-product-compliance
Lightning Source LLC
Chambersburg PA
CBHW082118230426
43671CB00015B/2728
* 9 7 8 1 7 8 0 3 9 0 0 8 6 *